XV Brazilian Symposium on High Performance Computational Systems (WSCAD 2014)

Journal of Physics: Conference Series Volume 649

Sao Jose dos Campos, Brazil
8 – 10 October 2014

Editors:

Alba Melo
Alvaro Fazenda
Denise Stringhini

ISBN: 978-1-5108-1476-9
ISSN: 1742-6588

Printed from e-media with permission by:

Curran Associates, Inc.
57 Morehouse Lane
Red Hook, NY 12571

Some format issues inherent in the e-media version may also appear in this print version.

Printed by Curran Associates, Inc. (2016)

For permission requests, please contact the Institute of Physics
at the address below.

Institute of Physics
Dirac House, Temple Back
Bristol BS1 6BE UK

Phone: 44 1 17 929 7481
Fax: 44 1 17 920 0979

techtracking@iop.org

Additional copies of this publication are available from:

Curran Associates, Inc.
57 Morehouse Lane
Red Hook, NY 12571 USA
Phone: 845-758-0400
Fax: 845-758-2634
Email: curran@proceedings.com
Web: www.proceedings.com

Table of contents

Volume 649

XV Brazilian Symposium on High Performance Computational Systems (WSCAD 2014)
8–10 October 2014, São Jose dos Campos, SP, Brazil

Accepted papers received: 1 October 2015
Published online: 19 October 2015

Preface

Papers

Journal of Physics: Conference Series **649** (2015) 011001 doi:10.1088/1742-6596/649/1/011001

Journal of Physics: Conference Series - Special issue for WSCAD 2014 - XV Brazilian Symposium on High Performance Computational Systems

Alba Melo[1], Alvaro Fazenda[2] and Denise Stringhini[2]

[1]**Brasilia University (UNB), Brasilia, DF, Brazil**

email: albamm@cic.unb.br

[2]**Federal University of Sao Paulo (ICT-UNIFESP), Sao Jose dos Campos, SP, Brazil**
email:{alvaro.fazenda,dstringhini}@unifesp.br

Preface

We are very pleased to welcome you to this edition of the Journal of Physics: Conference Series. This is a special issue for WSCAD 2014 (XV Brazilian Symposium on High Performance Computational Systems) which comprises a set of seven papers very carefully selected from the best papers from the WSCAD conference in 2014. The authors of the selected papers submitted an extended version of their work and then all the papers went through a new review process. We are thankful to the authors for their contributions to this special issue. Next, a brief introduction to these papers.

The first four papers are concerned with the use of accelerators and quantum computing simulation. The first paper is "LALPC: Exploiting Parallelism from FPGAs Using C Language". It presents LALPC, a prototype high-level synthesis tool, specialized for hardware generation for loop-intensive code segments. LALPC extends the functionality of a previous tool by using a subset of the C language as input code to describe computations, improving the usability and potential acceptance of the technique among developers. LALPC also enhances parallelism exploitation by applying loop unrolling, and providing support for automatic generation and scheduling of parallel memory accesses.

Next, the second paper "A mechanism to reduce energy waste in the post-execution of GPU applications" is focused on post-execution energy savings. When the post-execution behavior is analyzed in newer GPU ("Graphics Processing Unit") cards, it is observed that the power draw does not return to the idle state in an efficient way, creating an unexpected power wastage. To overcome the power wastage and inefficient returning to idle in the post-execution, the authors developed a strategy to reduce the energy consumption considering a minimal impact on global performance.

The third paper is "Code optimization by using a GPU applied to a Dataflow numerical simulation model". It aims to study techniques for parallel computing using a GPU in order to optimize the performance of a fragment of computational code, implemented as a data-flow system which is part of a meteorological numerical model responsible for calculating the advection transportation phenomena. The possible algorithm limitations for GPU efficiency were also addressed.

The fourth paper is "Optimizing Quantum Simulation for Heterogeneous Computing: a Hadamard Transformation Study". It presents the contributions in the optimization of the environment VirD-GM, conceived in three steps: (i) the theoretical studies and implementation of the abstractions of the Mixed Partial Process defined in the qGM model, focusing on the reduction of the memory consumption regarding multidimensional QTs; (ii) the distributed/parallel implementation of such abstractions allowing its execution on clusters of GPUs; (iii) and optimizations that predict multiplications by zero-value of the quantum states/transformations, implying reduction in the number of computations.

Next, we present the final sequence with three papers focusing on Cloud Computing and Scheduling. The fifth paper is "Middleware for Processing Message Queues with Elasticity Support and Sequential Integrity of Asynchronous Message Processing". It proposes a middleware solution to dynamically analyze the flow of message queues, and a mechanism to increase the parallelized consumption based on the output behavior. An architecture for IOD (Increase On Demand) middleware is presented, with support for the increase and decrease of threads to cope with the growth of message queues, using the technique of limit-based heuristics over a given period of time and grouping messages into sub-queues based on classification criteria.

The sixth paper "Towards Cloud-based Asynchronous Elasticity for Iterative HPC Applications" proposes a PaaS-based model called AutoElastics aimed at providing cloud elasticity for HPC applications in an efficient and transparent manner. In particularl, AutoElastic is focusing on master-slave iterative applications, but offers a flexible framework to support other HPC programming styles, such as pipeline and BSP.

Finally, the paper "ComprehensiveBench: a Benchmark for the Extensive Evaluation of Global Scheduling Algorithms" presents a benchmark for global scheduling algorithms that enables the variation of a vast range of parameters that affect performance. ComprehensiveBench can be used to assist in the development and evaluation of new scheduling algorithms, to help choose a specific algorithm for an arbitrary application, to emulate other applications, and to enable statistical tests.

The guest editors would like to thank all the program committee members, session chairs and the audience of the WSCAD 2014 for their work and participation. We also would like to thank the editors of the JPCS for the opportunity, the Editorial Board members and especially the journal invited reviewers, for their important collaboration, reviewing the final papers that allowed this special issue become a reality.

Invited reviewers

We would like to thank the invited reviewers and the high quality of their work:

- Adenauer Correa Yamin (adenauer@inf.ufpel.edu.br), UFPEL
- Alexandro Baldassin (alex@rc.unesp.br), UNESP

- Andrea Charao (andrea@inf.ufsm.br), UFSM
- Antonio Roberto Mury (a.roberto.m@gmail.com), LNCC
- Arlindo Flavio Da Concei¸c˜ao (arlindoflavio@gmail.com), UNIFESP
- Calebe de Paula Bianchini (prof.calebe.bianchini@gmail.com), Mackenzie
- Carla Osthoff Ferreira de Barros (carla.osthoff@gmail.com), LNCC
- Celso L. Mendes (cmendes@illinois.edu), University of Illinois
- Daniel Cordeiro (danielc@ime.usp.br), USP
- Esteban Meneses (emeneses@pitt.edu), University of Pittsburgh
- Gabriel Silva (gabriel@dcc.ufrj.br), UFRJ
- Henrique Cota de Freitas (cota@pucminas.br), PUC Minas
- Luciana Arantes (Luciana.Arantes@lip6.fr), Universite Pierre et Marie Curie - Paris 6
- Marco Netto (marco.netto@gmail.com), IBM Research
- Pedro Diniz (pedro@isi.edu), USC/ISI
- Paulo Sergio Lopes de Souza (pssouza@icmc.usp.br), ICMC/USP
- Rafael Ramos Dos Santos (rafael.santos@ceitec-sa.com), CEITEC SA
- Raphael Yokoingawa de Camargo (raphael.camargo@ufabc.edu.br), UFABC
- Roberto Pinto Souto (rpsouto@lncc.br), LNCC
- Rodolfo Jardim Azevedo (rodolfo@ic.unicamp.br), UNICAMP
- Siang Song (siangsong@gmail.com), UFABC and USP
- Tatiana Gadelha Serra Dos Santos (tatiana.santos@ceitec-sa.com), CEITEC
- Tiago Coelho Ferreto (tiago.ferreto@pucrs.br), PUCRS
- Vinod Rebello (vinod@ic.uff.br), UFF

Editors' biographical notes

Alba Cristina M. A. Melo received her Ph.D in Computer Science from the Institut National Polytechnique de Grenoble (INPG), France, in 1996, her M.Sc. in Computer Science from Federal University of Rio Grande do Sul, Brazil, in 1991 and her B.Sc. in Computer Science from University of Brasilia, Brazil, in 1986. She is currently an Associate Professor in the Computer Science Department at the University of Brasilia, Brazil. She is a Senior Member of the IEEE Computer Society and she is currently a CNPq/Brazil Research Fellow level 1D. Her research interests include high performance computing, cloud computing and computational biology.

Álvaro L. Fazenda received the M.Sc. and Ph.D. degrees in applied computing, both from the National Institute for Space Research (INPE), Brazil, in 1997 and 2002, respectively. He is an Assistant Professor at the Federal University of Sao Paulo (Unifesp), Brazil. His research interests include high performance computing and scientific computing.

Denise Stringhini received her degree in Bachelor in Informatics from Pontifícia Universidade Católica do Rio Grande do Sul in 1993, and obtained her M.Sc. and Ph.D degrees in Computer Science both from Federal University of Rio Grande do Sul, Brazil, in 1997 and 2002, respectively. Currently she is Assistant Professor at Federal University of Sao Paulo (Unifesp), Brazil. Her research interests include high performance computing, parallel programming and tools, clusters and accelerators.

Peer review statement

All papers published in this volume of *Journal of Physics: Conference Series* have been peer reviewed through processes administered by the proceedings Editors. Reviews were conducted by expert referees to the professional and scientific standards expected of a proceedings journal published by IOP Publishing.

LALPC: Exploiting Parallelism from FPGAs Using C Language

Lucas F. Porto[1], Marcio M. Fernandes[1], Vanderlei Bonato[2], Ricardo Menotti[1]

[1] Department of Computer Science
UFSCAR - Federal University of São Carlos
São Carlos - SP, Brazil
[2] Institute of Mathematical and Computer Sciences
USP - University of São Paulo
São Carlos - SP, Brazil

E-mail: `menotti@dc.ufscar.br`

Abstract. This paper presents LALPC, a prototype high-level synthesis tool, specialized in hardware generation for loop-intensive code segments. As demonstrated in a previous work, the underlying hardware components target by LALPC are highly specialized for loop pipeline execution, resulting in efficient implementations, both in terms of performance and resources usage (silicon area). LALPC extends the functionality of a previous tool by using a subset of the C language as input code to describe computations, improving the usability and potential acceptance of the technique among developers. LALPC also enhances parallelism exploitation by applying loop unrolling, and providing support for automatic generation and scheduling of parallel memory accesses. The *combination* of using the C language to automate the process of hardware design, with an efficient underlying scheme to support loop pipelining, constitutes the main goal and contribution of the work described in this paper. Experimental results have shown the effectiveness of those techniques to enhance performance, and also exemplifies how some of the LALPC compiler features may support performance-resources trade-off analysis tasks.

1. Introduction

Nowadays, reconfigurable computing, in particular modern FPGAs (*Field-Programmable Gate Arrays*), can be considered a good platform for the development of highly parallel systems with low power consumption. Many hybrid architectures adopt software to execute large code sections, and hardware accelerators to run performance critical sections of a given application. Due to software code reuse, this approach is able to increase performance while improving productivity in the development process [1].

In embedded systems, the hardware/software combination is usually tightly coupled, either using a reconfigurable soft-core processor in the device itself, or via devices and kits that combine FPGAs and a processor. In this context, it is worth mentioning some contemporary solutions such as Xilinx Zynq FPGA family [2], which includes ARM cores hardwired, and Altera/Intel DE2i-150 kit [3], which interconnects an Atom processor with a Cyclone IV FPGA through a PCIe bus. Several recent studies report the use of FPGAs to accelerate applications, such as [4], and many others presented in [5].

Despite the potential for high performance gains and power savings, reconfigurable computing is still considered by many designers as a hardware solution requiring long development times. This limitation is of particular concern for the adoption of hybrid architectures relying on FPGA accelerators. HLS (*High-Level Synthesis*) tools – used to convert more abstract descriptions directly into circuits – have been proposed since the 90s, in an attempt to facilitate hardware development, bringing it closer to software developers. After more than 20 years of steady progress, more mature tools have emerged, and those have become almost a necessity for the following reasons: i) need for rapid design space exploration from functional specifications; ii) high complexity of current devices, requiring a higher level of abstraction for large projects; iii) need for improved productivity by means of components reusing; iv) need for accurate system-level verification; v) increasing adoption of accelerators and heterogeneous SoCs (*System-on-Chips*) [6].

Several challenges face the design of a HLS tool. Besides the goal of delivering RTL (*Register-Transfer Level*) circuits as efficient as possible regarding aspects such as area, performance, and energy consumption, the input language is determinant for the tool adoption and effective use. One of the approaches presented in Section 2 is LALP (*Language for Aggressive Loop Pipelining*), a scheme aimed to generate hardware accelerators targeting the optimization of software loops. The technique relies on a special purpose language, and underlying hardware model specially designed to optimize the execution of pipelined loops [7].

Experimental results using LALP techniques have shown promising results in terms of time and area, which has motivated further developments to improve the way computations are specified by a system's designer: instead of using the special purpose language, the framework has been extended to use the C Language, allowing for easier translation of software algorithms into hardware structures. The use of C language is a desirables feature as most algorithms of interest for embedded systems have well tested reference implementations described in that language. The LALPC (*C-based Language for Aggressive Loop Pipelining*) compiler infrastructure presented in this paper extends the functionality of LALP not only in terms of the input language, but also in the strategies used to increase parallelism exploitation [8]. Besides loop pipelining, automatic loop unrolling and parallel memory access are extensively used whenever possible, and thus improving overall performance results. By doing so, LALPC improves the tool usability and acceptance through the use of C language, while still retaining the area and time advantages of the underlying hardware scheme supporting loop pipelining. The *combination* of those two features constitutes the main contribution of this work, which has shown promising results when compared against some current similar state-of-the-art tools.

The remaining sections of this paper are organized as follows: Section 2 presents some related works. In Section 3 the LALPC compiler infrastructure and its main features are described, followed by some experimental results in Section 4. Section 5 brings some final remarks regarding the work presented in this paper.

2. Related Work

Several high-level synthesis tools have been proposed recently using various approaches, which shows the growing interest in automatic generation of hardware for reconfigurable devices [9]. Among others, some notable academic works are Spark [10], Haydn-C [11], ROCCC [12], and MATISSE [13]. Commercial tools have also matured, and a number of them are currently offered, such as Altera SDK for OpenCL, Xilinx Vivado, Calypto Catapult, Synopsys Synphony C, Cadence C-to-Silicon, ImpulseC and Cynthesizer. Metamodeling has also being exploited for system level design [14], where hardware and software are represented in a common environment. Such approach allows a hardware synthesis tool to use several information from the entire system in order to meet requirements.

An interesting alternative is ReflectC, a high-level synthesis tool with a focus on improving flexibility during the design process [15]. ReflectC can be used in conjunction with LARA [16], a special purpose aspect-oriented programming language. LARA has been designed for code instrumentation, controlling the application of code transformations and compiler optimizations. The language allows developers to capture non-functional requirements from applications in a structured way, while still keeping the original application source-code. By doing so, it enables effective application and control of the different tools available in the high-level synthesis toolchain, taking into account domain and target-specific aspects. This strategy improves the efficiency and coverage of the design-space exploration process, and ultimately may expose additional opportunities for hardware-software partitioning.

LegUp [17] is an open source project that has received considerable attention from the scientific community. The tool focus is on high-level synthesis using the C language as input. The C code is automatically compiled, and corresponding hardware structures are generated, described in Verilog HDL (*Hardware Description Language*). LegUp uses as front-end the LLVM compiler [18], which also performs the initial steps of analysis and optimization for further synthesis. LegUp implements a new back-end in order to target a hardware platform [17]. The tool can synthesize a substantial part of the C language, supporting various structure types, such as multidimensional arrays, global variables, pointers, and even floating-point values. The hardware generated by the LegUp tool has comparable quality to those generated by some available commercial tools for high-level synthesis [19]. In addition, it allows the integration of various accelerators with programs described in OpenMP and Pthreads [20].

The LALP compiler adopts a focused approach to high-level synthesis, aiming to optimize algorithms relying on time critical loops [7]. The basis of this approach is called ALP (*Aggressive Loop Pipelining*), a hardware scheme which adapts some of the ideas proposed in [21] for execution in FPGA platforms. However, instead of a data-driven scheme, the ALP technique attempts to achieve the maximum throughput by using counters to furnish the iteration space in loops, and shift-registers to synchronize operations [22]. By doing so, it avoids the need of a centralized control based on finite state machines, a *key difference* from traditional approaches to loop pipelining. In the ALP scheme, operations in the loop body are executed at clock cycles according to the paths taken during execution, and optimal loop pipelining can be achieved, without performance losses due to an FSM controller.

The LALP tool allows design space exploration through marking directives in the code. These directives are used to control the exact clock cycle of operations, generating different architectures for the same input code. Another important feature is that the LALP compiler uses a specific language for describing the algorithm. The LALP language has a higher level of abstraction when compared to VHDL and Verilog languages. The results obtained for this domain may be very efficient, as demonstrated in [23], however, the use of a *special purpose language* limits its use since it implies a learning curve for any developer wishing to use it.

Considering that the LALPC tool presented in this paper was built upon ALP, a specialized hardware scheme offering time and area gains for loop execution over some other existing tools [22], it can be argued that LALPC keeps this advantage over similar HLS tools. The support for C language input extends LALPC applicability, and so its chances of outperforming existing HLS tools for loop optimizations.

3. LALPC Compiler

The LALPC compiler presented in this paper extends LALP to support the C language as input, and also employs new strategies for parallelism exploitation. As already pointed out, one of the main changes in relation to the LALP compiler is the use of C language as input for describing computations. This has been accomplished by using the ROSE compiler [24] as part of the overall LALPC compilation flow (Figure 1).

As a prototype implementation, LALPC allows for a limited subset of C language, although sufficient for testing purposes with representative application codes. In addition to sequential constructs, it is possible to use the following constructions of the C language: regular (for) loops, block and ternary conditional checks, arithmetic and logic operations, accumulators, arrays, bit shifts. Some notable limitations include float data types, function calls, structs, pointers, multi-dimensional arrays, and recursion. Such limitations are due to the current capabilities of the LALPC back-end, some of them usually found in other HLS tools (e.g. float data types, function calls).

The LALPC compiler targets the same VHDL components library used in LALP [25]. This hardware library is optimized for *loop pipelining* execution, a performance oriented feature at the core of the system. The generated VHDL code can be synthesized both in Altera and Xilinx tools.

The compiler allows using markup directives to guide the tool during the hardware generation process. These markup directives are described as *pragmas* inserted in the code, which are them handled by the compiler. By varying some parameters through the use of pragmas, it is possible to generate different architectures with specific features for the same input source code. With pragmas, the programmer can apply hardware customizations based on knowledge from a high abstraction level of implementation details. Pragmas also controls the use of other features aiming to improve parallelism exploitation, in particular *loop unrolling*, and the use of *multi-port memories* for simultaneous array accesses. The use of pragmas in LALPC compiler is described in details in Section 3.2.

The previous LALP tool required the programmer to use directives to control the timing of operations. This manual synchronization ensures that the scheduling process will always run properly. Although the LALPC compiler still supports the use of manual synchronization (by using pragmas), it is no longer a requirement. By relying on the more robust analysis passes and intermediate representation provided by the ROSE compiler, it is possible for the framework to infer valid synchronized sequences of operations. All tests conducted in this study showed the correctness of this automatic approach, however, it does not implies on a formal compiler validation at this stage.

3.1. Compiler Structure

The LALPC instantiates parts of the ROSE compiler to validate the input code in C language and to apply some optimization techniques. After this step, an intermediate representation, built in the form of an AST (Abstract Syntax Tree) with the intermediate representation of the code is obtained. This representation is used to generate the necessary hardware components, and to define the data and control connections among them.

The ROSE compiler is capable of performing a number of analyses and optimizations in the input code. Of particular interest is the code representation in Static Single Assignment (SSA) form, making unique every assignment to a variable. This optimization eases the hardware generation process, a step manually performed for the previous compiler.

Some pragmas are handled directly by the ROSE compiler front-end, which are reflected in the AST representation (e.g. loop unrolling). Others are dealt with by the hardware-oriented back-end (e.g. operations bit-width). After creating the signals among components, the compiler performs the scheduling and balancing steps [26]. Finally, the VHDL compiler generates files with the components and their connections. Figure 1 shows the structure of the compiler, which also outputs auxiliary internal compiler representations for a visualization tool (Graphviz).

3.2. Pragmas

High-level synthesis tools (both research and commercially oriented) usually have limitations arising from differences between the paradigms of software and hardware development. Software programming languages aim to map a given problem in a well-defined general-purpose

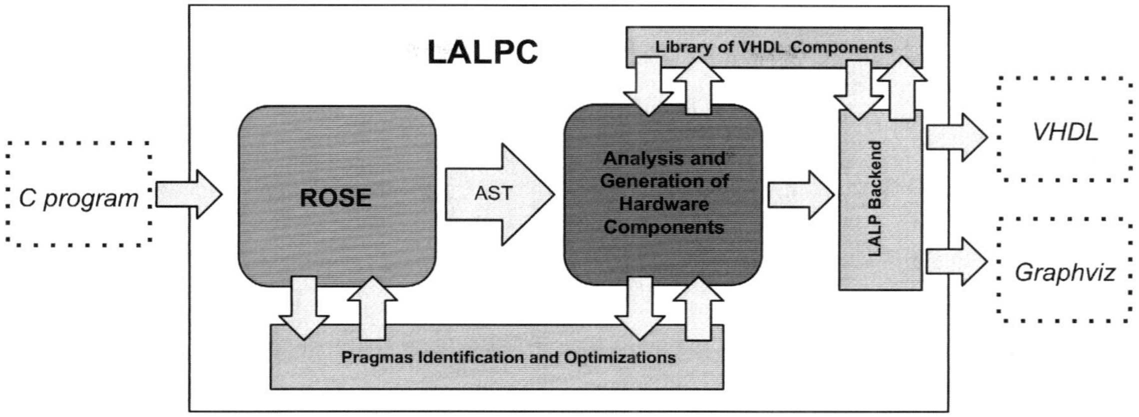

Figure 1: The LALPC Compiler Structure

architecture. The construction of a custom hardware architecture for a given problem may take into consideration different parameters, since there is no fixed bounding to previously defined functional units, memory hierarchy and so on. These differences may lead the synthesis tool to generate architectures that require a large amount of hardware resources. Code restructuring and/or fine tuning of hardware generation parameters is often a strategy for mitigating undesirable effects.

Programming directives via pragmas is intended to allow the programmer to define some specific characteristics during the build process. The use of pragmas modifies the hardware generated by the compiler, which can result in different architectures. For the same input code, not only different performance levels may be obtained, but significant variations in resources usage and power consumption in the adopted reconfigurable device. The use of pragmas is also justified by the existing features and limitations in C when used for hardware specification.

To represent the different paradigms between the software described in C language and hardware architectures, consider, for example, the need to specify a function with multiple outputs. In the LALPC compiler it is possible to set pragmas that indicates which variables may have output pins in their respective registers.

The pragmas currently treated by the LALPC compiler are described below:

- `#pragma alp bit` – defines a boolean (single bit) register, since the C language does not have a data type for this. This pragma is useful to define control signals, preventing wider variables being used for this purpose.
- `#pragma alp data_width` – defines the bit width used by the VHDL components. This program can be used to increase or decrease the field of representation of numbers according to the need, providing a space-saving mechanism. The VHDL component library is fully parameterized in terms of bit width, the compiler can then instantiate components with the width specified in this pragma.
- `#pragma alp delay` – controls clock cycles of hardware components, allowing manual synchronization of operations. This feature can be used to improve the circuit's operating frequency (e.g. putting additional registers), which enables the retiming in synthesis tool.
- `#pragma alp multiport` – checks for multiple memory accesses, and generates custom multiport RAM modules. This feature is explained in more detail below.
- `#pragma alp out` – creates output pins in VHDL components, typically used to facilitate data analysis and the generation of blocks with multiple outputs. The execution of a

function in software is temporal, resulting in a value at the end of all steps. In hardware – with spatial execution in pipeline – a block can simultaneously calculate several iterations. This pragma can then be used to extract intermediate values of the computation, which can be useful in other parts of the algorithm.

- `#pragma alp unroll` – used to apply the loop unrolling optimization, resulting in performance gains for application codes based on loop-critical sections, as can be seen below. The handling of this pragma is delegated to the ROSE compiler, invoking a method that works directly in its intermediate representation.

3.3. Parallelism Exploitation

The *multiport* and *unroll pragmas* are used to apply optimization techniques to speed up the execution speed of applications. However, its use may also affects the amount of resources allocated in the reconfigurable device, as can be seen below. For this reason, a level of design space exploration may be required to balance performance and resources utilization.

3.3.1. ***#pragma alp multiport*** Most of today's FPGAs have dual port memories internally. This pragma generates a RAM component with multiple ports, allowing simultaneous read access. Write parallel access are limited to a factor of two. This improves the final performance of the application in many cases: serial array access only constrained by memory ports can be overcame by the use of multiport memory banks, allowing for parallel access. The generation of these memories is delegated to the synthesis tool and is beyond the scope of this paper. To have more control over the process that could be treated with techniques such as those described in the literature [27, 28, 29].

3.3.2. ***#pragma alp unroll*** The use of the `#pragma alp unroll` instructs the compiler to analyse the operations within the loop, trying to identify dependencies between them. If dependencies permit, the compiler duplicates internal components of the loop, enabling more than one iteration to be executed in parallel. The result is a significant gain in the application processing when compared with the sequential version. On the other hand, the process of replicating internal operations increases the amount of used resources in the reconfigurable device, a growth proportional to the number of operations contained in the loop.

As an example, Figure 2a aims to exemplify the use of this pragma: line 4 specifies that, if possible, an unroll factor of two should be applied to the loop body. Figure 2b shows the resulting code generated by the ROSE compiler (in high-level language, for clarity purposes). This new structure becomes the intermediate representation that will be used by the LALPC back-end, and correponding generation of hardware components.

```
1 void vecsum() {
2   int i, x[N], y[N], z[N];
3   #pragma alp multiport
4   #pragma alp unroll 2
5   for (i = 0; i < N; i++)
6     z[i] = x[i] + y[i];
7 }
```

(a) Initial code.

```
void vecsum() {
  int i, x[N], y[N], z[N];
  for (i = 0; i < N; i+=2) {
    z[i]   = x[i]   + y[i];
    z[i+1] = x[i+1] + y[i+1];
  }
}
```

(b) Resulting code.

Figure 2: Example of *loop unrolling* provided by ROSE compiler

It can be seen that line 5 was generated from line 4, but with different access to the memories, which allows parallel processing of two iterations for each clock cycle.

Also in this example, when applying the unroll pragma the compiler generates more simultaneous accesses to the memories, being represented by the indexes "i" and "i+1" of the vectors. In this way it is possible to combine the pragma multiport for simultaneous access to this memory. Line 3 on Figure 2a contains the pragma that indicates to the LALPC compiler customize the memories when simultaneous accesses are found, regardless of reading or writing.

The combination of the two above pragmas increases significantly the performance in many applications, compared to the sequential processing. It is worth to note that the use of such features is an option when the developer needs to meet project requirements. Section 4 presents some experimental results showing the impact of using these pragmas in terms of performance and resources usage.

4. Experimental Results

Some experimental results were conducted in order to evaluate the hardware generated by the LALPC compiler. This work focused on performance and resources usage, comparing the results against the previous LALP compiler, and also against the LegUp compiler. Although all selected benchmarks rely heavily on loops, they present different characteristics in terms of size of the loop body, pattern and frequency of memory access, and data dependences. All benchmarks use 32-bit integers, unless stated. The benchmarks used for this work are:

- Accumulator algorithm [17] - 10 loop iterations;
- ADPCM encoder/decoder audio algorithms [30] - 1024 loop iterations;
- Sobel image processing algorithm [31] - 78 loop iterations;
- Dotprod, Max and Vecsum algorithms used in DSPs [32] - 2048 loop iterations;

The synthesis tool used was Quartus II 13.0.0 Web Edition (64-bit) targeting Altera EP4CGX150DF31C7 Cyclone IV GX Family reconfigurable device (FPGA). Terasic DE2i-150 board was used as a platform both for software and hardware tests. Software results were adopted as a baseline implementation, a reasonable choice when porting software implementations to hardware platforms. The software execution was performed on a Intel Atom Dual Core Processor N2600, with a 1MB cache and running at 1.6GHz. The adoption of the Atom microprocessor was due to its tight integration to the FPGA development board used for the hardware implementations presented in this work. Considering that this processor is offered as part of a commercial hardware-software hybrid solution, it can be argued that it is a representative microprocessor of a hypothetical software-only embedded system. In all tests performed for this work, only implementation with one thread of execution were employed.

Software performance figures were obtained with *perf-stat* tool running on Yocto 3.0.32 and Poky 1.3.2 reference system which is based on GCC 4.7.2. The raw data obtained for performance and hardware resources usage are showed in Table 1, and further discussed in the next subsections.

4.1. Performance

The data presented in Figure 3 compares the normalized execution times for all benchmarks running on three hardware architectures and on one microprocessor for the software implementation. As expected LALPC and LALP results are very similar, as they target the same underlying hardware components, highly specialized to loop pipelining execution. This is also the very reason why LALPC outperforms LegUp for the selected benchmarks. However, this advantage should not be expected in general implementations.

Journal of Physics: Conference Series **649** (2015) 012001 doi:10.1088/1742-6596/649/1/012001

Table 1: Hardware resources usage and execution time for software platform (Intel Atom), and hardware implementations generated by LegUp, LALP and LALPC.

Benchmark	**Platform Compiler**	**Logic Elements**	**Comb**	**Reg**	**Memory (Bits)**	**Execution Time(us)**
Accumulator	Software	–	–	–	–	13.31
	LEGUP	658	564	541	640	0.44
	LALP	163	109	138	512	0.11
	LALPC	159	109	139	512	0.10
ADPCM Coder	Software	–	–	–	–	17.48
	LEGUP	1067	995	661	36128	172.10
	LALP	800	638	37	40960	79.72
	LALPC	986	673	837	41460	119.30
ADPCM Decoder	Software	–	–	–	–	16.90
	LEGUP	1049	913	683	52512	95.97
	LALP	507	388	464	41090	16.61
	LALPC	628	420	574	41248	39.00
Dotprod	Software	–	–	–	–	43.79
	LEGUP	852	696	590	131072	152,09
	LALP	95	81	65	131072	28.92
	LALPC	118	102	82	131072	28.50
Max	Software	–	–	–	–	38.55
	LegUP	294	272	234	65356	31.46
	LALP	66	47	62	65356	12.79
	LALPC	94	75	84	65356	12.89
Sobel	Software	–	–	–	–	24.06
	LEGUP	1266	1156	820	6400	4.16
	LALP	797	683	504	8269	5.67
	LALPC	1158	754	991	8435	4.40
Vecsum	Software	–	–	–	–	43.92
	LEGUP	891	788	551	196608	117.85
	LALP	101	53	67	196608	18.95
	LALPC	124	74	83	196608	17.20

Although for most benchmarks there are modest performance gains when moving from LALP to LALPC, it is worth noting that this can be accomplished with much less development effort, since algorithms can be coded using C language in LALPC, without the need of explicit synchronization parameters. In LALP, ADPCM implementations were modified to achieve greater parallelism, severely altering the order of operations in the code. In LALPC version we prefer to keep the unmodified code to make a fair comparison with LegUp.

As seen in Table 1, most hardware implementations result in clearly better performance when compared to the software platform. The noticeable exception is the ADPCM coder algorithm. That is explained by the algorithm characteristics, which have data dependences preventing the use of loop pipeline optimizations. That results in a quasi-sequential execution at a much lower clock rate in the hardware platform, when compared to the 1.6 GHz Atom processor. The obtained results for the software implementations are the mean value based on five runs for each benchmark, presented within a 95% confidence interval.

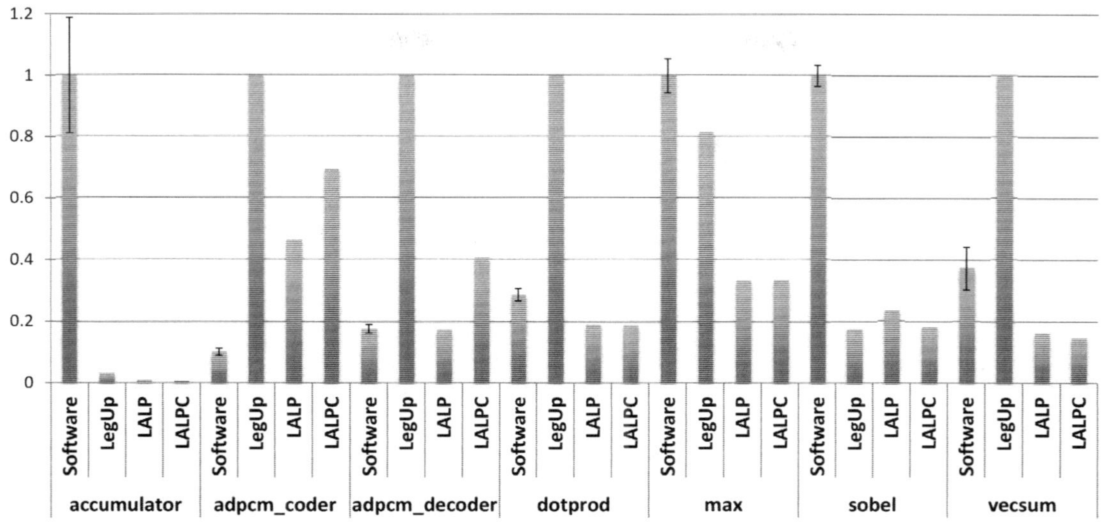

Figure 3: Comparison between normalized execution times for all implementations.

4.2. Speedup

For four benchmarks, the data in Table 2 shows the results obtained using optimization pragmas in terms of execution time and speedup, comparing all there compilers: LALPC, LALP, and LegUp. LALPC acheived an average speedup over LALP of 4.7x, and 10.6x when compared to LegUp. Those results demonstrated the effectiveness of the optimization techniques implemented in this work, for the selected benchmarks. However, the performance gain is usually obtained at the expense of additional hardware resources, as will be shown in the next sections. This observation leads to the need of trade-off analysis, shown in Sections 4.4 and 4.5.

Table 2: Speedup obtained using optimization pragmas.

Benchmark	Time LALPC (us)	Time LALP (us)	Speedup	Time LegUp (us)	Speedup
Accumulator	0.05	0.11	**2.2**	0.44	**8.8**
Dotprod	3.95	28.92	**7.3**	152.09	**38.5**
Sobel	0.45	5.67	**12.6**	4.16	**9.2**
Vecsum	9.73	18.95	**1.9**	117.85	**12.1**

4.3. Hardware Resources Usage

It can be observed in Table 1 that, in general, LegUp architectures requires more logic elements as it relies more heavily on Finite State Machines (FSMs), as opposed to LALPC and LALP. As for internal memories, most resources usage are similar, apart from those for the accumulator benchmark, which employs flip-flops instead of memories, in LALPC and LALP. Those two compilers always generate memory capacity (in number of words) as a power of 2, while LegUp is able to generate memories of arbitrary sizes. That can be clearly observed for the Sobel benchmark. LegUp instantiates memories using an IP core from Altera, while LALPC and LALP generates custom memories in VHDL.

As shown in Figure 4 (using a logaritmic scale) hardware implementations generated by LALPC are similar to those generated by LALP, in terms of hardware resources usage. This aspect was evaluated in terms of number of logical elements, combinational functions, logic registers, and memory bits. However, the results for both compilers are not identical, explained by the use of a new front-end and additional code transformations performed by the ROSE compiler. As already observed for the performance results, LALPC compares favourably against LegUp for the benchmarks used in this study. The most likely reason is again the specialized hardware components targeted by LALPC, an effect unlikely to appear for application codes not relying on loops.

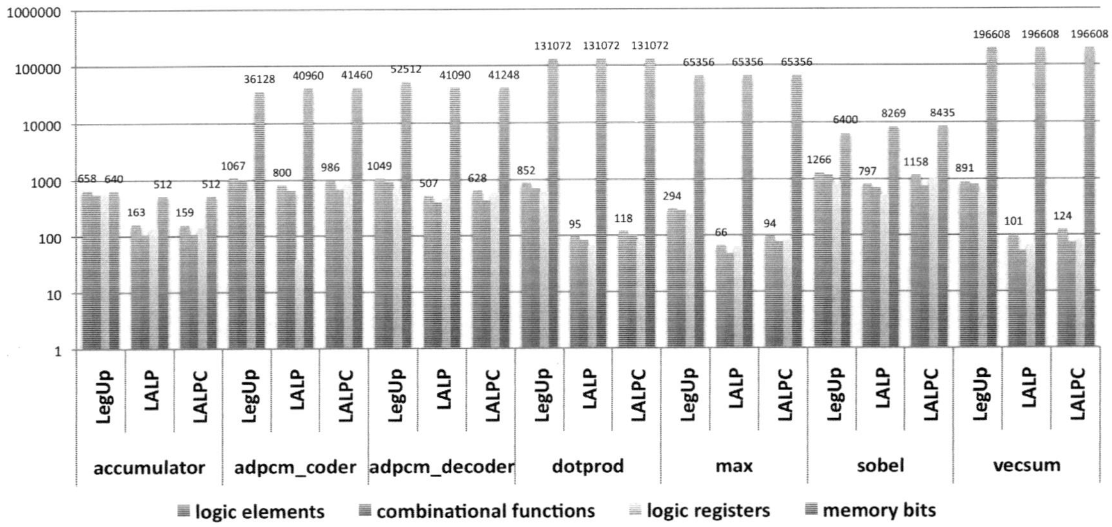

Figure 4: Hardware resources usage.

4.4. Design Space Exploration: Sobel

The use of pragmas has been already mentioned as a mechanism to help the designer in the task of fine tuning the compiler output, allowing for an easy exploration of some solution points trying to find an adequate balance between performance and hardware resources usage. In order to demonstrate the effective of this LALPC compiler feature, the Sobel benchmark was used. The Sobel benchmark was chosen due to some of its characteristics, it permits tests with the pragmas separately. One of the characteristics of interest of the algorithm is that it performs, for each loop iteration, 8 read accesses in the input vector, and a single write in the output vector after those data are processed. A number of hardware implementations were generated for the same input code, only varying the use of pragmas to control the level of optimizations employed.

In order to perform 8 read access to the RAM memory, a multiplexer is used to schedule the access to distinct addresses, making the reading step fully sequential. This process requires 8 cycles to read the pixels from memory for each iteration of the repeating loop. By using the *pragma multiport*, the compiler is able to replace the single port RAM memory module by a RAM memory with 8-read and 8-write ports. By using this customization, all read accesses are executed in a single cycle, and thus allowing a significant performance gain. However, this parallelism is achieve at high cost in terms of resources, since the actual implementation consists of 4 dual-port memories, with replicated input data.

Another optimization possible in this example is *loop unrolling*, because there are no dependencies within the repeating loop. By combining this optimization with the multiport pragma, the resulting hardware ensures a further performance gain, although at the expense of additional hardware resources. This combination doubles all existing operations within the loop kernel, generating 16 simultaneous accesses to the memory module containing the input image, and other two write accesses in the output memory. Unrolling a loop body increases the complexity of logic circuits as some of the data used in computations are moved from memory to internal registers. That causes a reduction in the number of employed memory bits, and an increase in the use of logic elements.

This performance-hardware trade-off can is graphically presented in Figure 5.

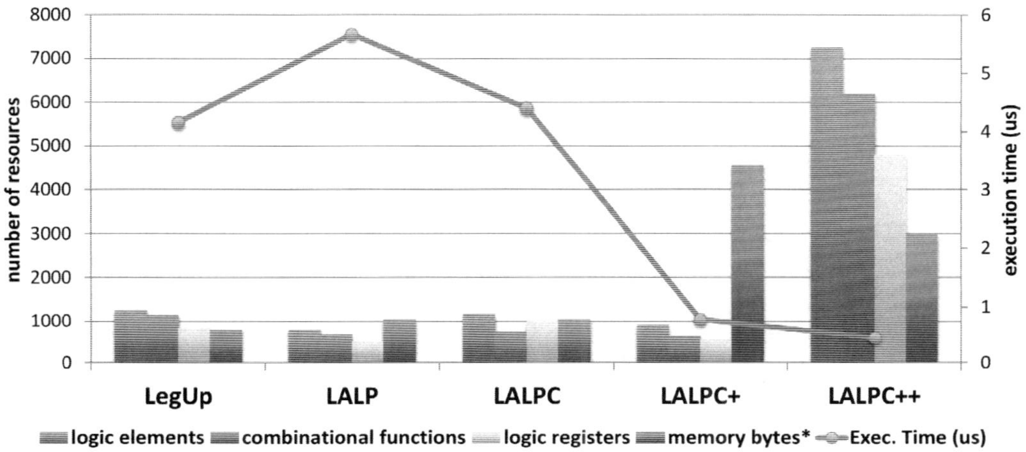

LALPC+: Multiport — LALPC++: Multiport & Unroll

Figure 5: Sobel Design Space Exploration: Resources vs Execution Time

Another evaluation easy performed by using the pragmas is the impact resulting from varying the memory and operations bit-width employed by the resulting architecture (Figure 6).

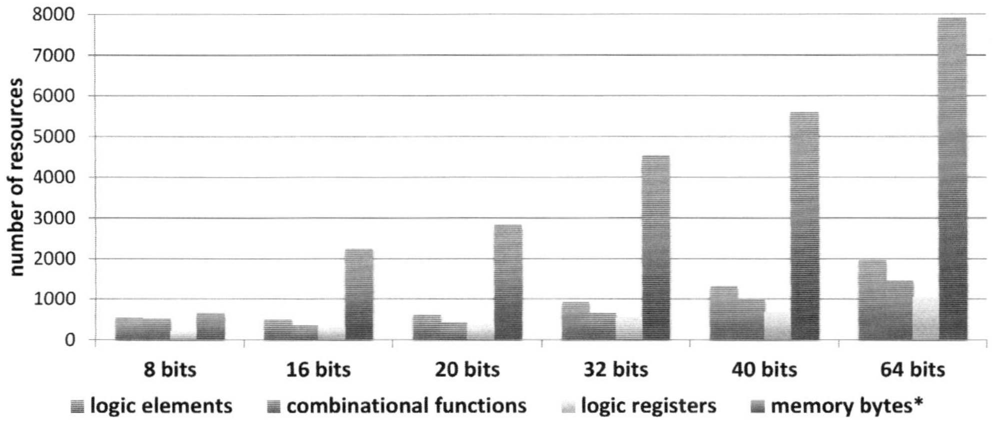

Figure 6: Sobel Design Space Exploration: Memory and Operations Bit-width

Those experiments exemplify the potential of the optimization techniques implemented in LALPC compiler. However, it is worth noticing that they also may affect the amount of resources required for the hardware synthesis, and could make it infeasible depending in case of limited resources in the reconfigurable device adopted for a given project.

4.5. Design Space Exploration: Accumulator, Vecsum, Dotprod

After completion of the initial tests for the evaluation of pragmas with the Sobel benchmark, new tests were performed with some of the other benchmarks mentioned above. However, applications that have intra-iteration loop dependences do not benefit from the unroll pragma. Because of that, some benchmarks of the list were not considered in this analysis. As shown in Figure 7, the same results pattern observe for Sobel were observed for the Accumulator, Vecsum, and Dotprod benchmarks: a steady improvement in performance resulting from the use of loop unrolling and parallel memory accesses, at the expense of additional hardware resources. Although this result is expected, it should be noticed that the LALPC compiler allows a quantitative evaluation of all design points with little modifications in the source code. The results for the accumulator benchmark show little variation in resources usage, but also little performance gains, which is due to a small number of loop iterations in the code. Execution time is halved for Vecsum, but it does not scale very well (exponential), as all data are moved to registers. Dotprod relies on resource replication but without increasing the use of memory, resulting in linear scaling for resources usage.

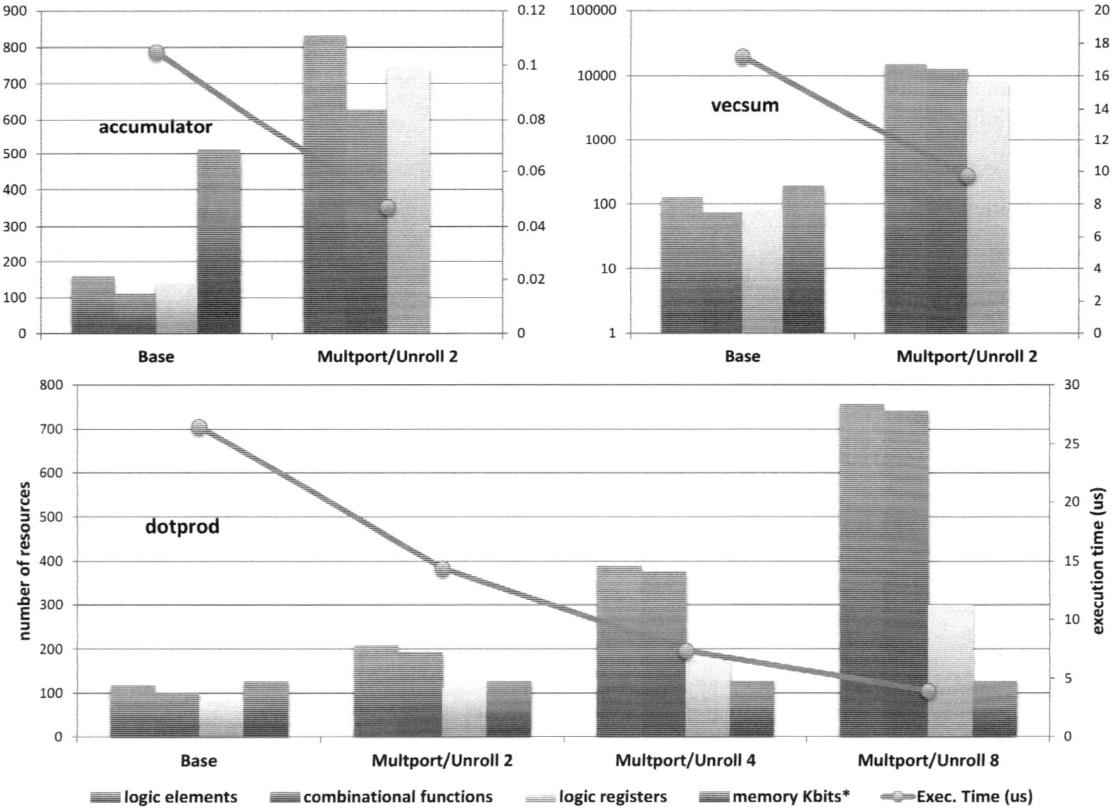

Figure 7: Accumulator, Vecsum, Dotprod Design Space Exploration: Resources vs Execution Time

5. Conclusion

This work has presented the LALPC compiler, a high-level synthesis tool, which was built as an extension of an existing tool (LALP). From the user's point of view, the main advantage in relation to the previous tool is the use of C language as input code to describe computations. LALPC also supports the use of pragmas to guide the hardware generation process, and to enable some optimizations such as loop unrolling, and parallel memory access. It has been demonstrated that those features can be useful support time-area trade-off analyses during the development process. The combination of using C language with an efficient hardware scheme supporting loop pipelining constitutes the main contribution of this work.

Although allowing only a subset of the full C language specification, the LALPC compiler was able to generate efficient implementations for some loop-oriented kernel benchmarks. Those results compare favourably against those obtained using the LegUp compiler, although we acknowledge this can be achieve only for loop intensive code segments. It is also worth noticing that LegUp supports a broader C language subset, which obviously increases its applicability when compared to LALPC. Nevertheless, the results obtained in this work suggests that the specialized techniques employed by LALPC have the potential to be further improved, or incorporated as a specialized feature targeting loops on other existing HLS tools.

References

[1] Liao C, Quinlan D J, Willcock J J and Panas T 2010 *International Journal of Parallel Programming* **38** 361–378

[2] Santarini M 2011 *Xcell J* **75** 8–13

[3] Terasic Technologies Inc. 2013 *DE2i-150 FPGA System User Manual* URL `http://www.terasic.com/`

[4] Putnam A, Caulfield A M, Chung E S, Chiou D, Constantinides K, Demme J, Esmaeilzadeh H, Fowers J, Gopal G P, Gray J, Haselman M, Hauck S, Heil S, Hormati A, Kim J Y, Lanka S, Larus J, Peterson E, Pope S, Smith A, Thong J, Xiao P Y and Burger D 2014 *Computer Architecture (ISCA), 2014 ACM/IEEE 41st International Symposium on* pp 13–24

[5] Vanderbauwhede W and Benkrid K 2013 *High-Performance Computing Using FPGAs* (Springer)

[6] Cong J, Liu B, Neuendorffer S, Noguera J, Vissers K and Zhang Z 2011 *Computer-Aided Design of Integrated Circuits and Systems, IEEE Transactions on* **30** 473–491 ISSN 0278-0070

[7] Menotti R, Cardoso J M P, Fernandes M M and Marques E 2012 *International Journal of Parallel Programming* **40** 262–289

[8] Porto L F and Menotti R 2014 *Anais da 15a edição do Simpósio em Sistemas Computacionais de Alto Desempenho* pp 87–98 ISSN 2358-6613

[9] Martin G and Smith G 2009 *IEEE Design and Test of Computers* **26** 18–25 ISSN 0740-7475

[10] Gupta S, Gupta R, Dutt N and Nicolau A 2004 *SPARK: A Parallelizing Approach to the High-Level Synthesis of Digital Circuits* (Kluwer Academic Publishers)

[11] Coutinho J G F and Luk W 2003 *Field-Programmable Technology (FPT), 2003. Proceedings. 2003 IEEE International Conference on* (IEEE) pp 278–285

[12] Buyukkurt B, Guo Z and Najjar W 2006 *Reconfigurable Computing: Architectures and Applications* 401–412

[13] Bispo J, Pinto P, Nobre R, Carvalho T, Cardoso J and Diniz P 2013 *Industrial Informatics (INDIN), 2013 11th IEEE International Conference on* pp 602–608

[14] Sangiovanni-Vincentelli A, Yang G, Shukla S, Mathaikutty D and Sztipanovits J 2009 *Design Test of Computers, IEEE* **26** 54–69 ISSN 0740-7475

[15] Cardoso J M, Diniz P C, Petrov Z, Bertels K, Hübner M, van Someren H, Gonçalves F, de Coutinho J G F, Constantinides G A, Olivier B *et al.* 2011 *Reconfigurable Computing* (Springer) pp 261–289

[16] Cardoso J M P, Carvalho T, Coutinho J G F, Nobre R, Nane R, Diniz P C, Petrov Z, Luk W and Bertels K 2013 *Microprocessors and Microsystems - Embedded Hardware Design* **37** 1073–1089 URL `http://dx.doi.org/10.1016/j.micpro.2013.06.001`

[17] Canis A, Choi J, Aldham M, Zhang V, Kammoona A, Anderson J H, Brown S and Czajkowski T 2011 *Proceedings of the 19th ACM/SIGDA international symposium on Field programmable gate arrays* (ACM) pp 33–36

[18] Lattner C and Adve V 2004 *Proceedings of the international symposium on Code generation and optimization: feedback-directed and runtime optimization* CGO '04 (Washington, DC, USA: IEEE Computer Society) pp 75– ISBN 0-7695-2102-9

[19] Canis A, Choi J, Aldham M, Zhang V, Kammoona A, Czajkowski T, Brown S D and Anderson J H 2013 *ACM Trans. Embed. Comput. Syst.* **13** 24:1–24:27 ISSN 1539-9087 URL `http://doi.acm.org/10.1145/2514740`

[20] Choi J, Brown S and Anderson J 2013 *Field-Programmable Technology (FPT), 2013 International Conference on* (IEEE) pp 270–277

[21] Cardoso J M P 2005 *CF '05: Proceedings of the 2nd conference on Computing Frontiers* (New York, NY, USA: ACM Press) pp 106–115 ISBN 1-59593-019-1

[22] Menotti R, Marques E and Cardoso J M P 2007 *International Conference on Field Programmable Logic and Applications (FPL)* pp 501–502

[23] Menotti R, Cardoso J M P, Fernandes M M and Marques E 2009 *Proceedings of the 21st International Symposium on Computer Architecture and High Performance Computing (SBAC-PAD)* (Los Alamitos, CA, USA: IEEE Computer Society Press) pp 3–10

[24] Quinlan D *et al.* 2012 Rose compiler infrastructure URL `http://rosecompiler.org/`

[25] Cardoso J M P 2000 *Compilação de Algoritmos em Java para Sistemas Computacionais Reconfiguráveis com Exploração do Paralelismo ao Nível das Operações* Ph.D. thesis Universidade Técnica de Lisboa

[26] Menotti R, Cardoso J M P, Fernandes M M and Marques E 2011 *International Journal of Parallel Programming* 1–28

[27] Malazgirt G, Yantir H, Yurdakul A and Niar S 2014 *Field Programmable Logic and Applications (FPL), 2014 24th International Conference on* pp 1–4

[28] Yantir H, Bayar S and Yurdakul A 2013 *Digital System Design (DSD), 2013 Euromicro Conference on* pp 185–192

[29] LaForest C E and Steffan J G 2010 *Proceedings of the 18th Annual ACM/SIGDA International Symposium on Field Programmable Gate Arrays* FPGA '10 (New York, NY, USA: ACM) pp 41–50 ISBN 978-1-60558-911-4 URL `http://doi.acm.org/10.1145/1723112.1723122`

[30] Guthaus M R, Ringenberg J S, Ernst D, Austin T M, Mudge T and Brown R B 2001 *WWC '01: Proceedings of the Workload Characterization, 2001. WWC-4. 2001 IEEE International Workshop* (Washington, DC, USA: IEEE Computer Society) pp 3–14 ISBN 0-7803-7315-4

[31] Texas Instruments Incorporated 2003 *TMS320C64x Image/Video Processing Library: Programmer's Reference* URL `http://focus.ti.com/lit/ug/spru023b/spru023b.pdf`

[32] Texas Instruments Incorporated 2003 *TMS320C64x DSP Library: Programmer's Reference* URL `http://focus.ti.com/lit/ug/spru565b/spru565b.pdf`

A mechanism to reduce energy waste in the post-execution of GPU applications

Emmanuell D. Carreño, Adiel S. Sarates Jr. and Philippe O. A. Navaux

Informatics Institute – Federal University of Rio Grande do Sul (UFRGS)
P.O Box 15.064 – 91501-970 – Porto Alegre – RS – Brazil

E-mail: {edcarreno,assaratesj,navaux}@inf.ufrgs.br

Abstract. With the increasing demand of GPU accelerators for general purpose in HPC, the impact of energy consumption of these resources cannot be overlooked. To reduce the power consumption some strategies have been applied, but their approaches have been mostly focused on power savings during the application execution. This work focuses on post-execution energy savings. When the post-execution behavior is analyzed in newer GPU cards, it is observed that the power draw does not return to the idle state in an efficient way, creating an unexpected power waste. To overcome this inefficient return to idle and power draw waste in the post-execution, we developed a strategy to reduce the energy consumption considering a minimal impact on global performance. Using this strategy, we achieved energy savings up to 73 percent in the post-execution phase of a single run of a GPU application. In the case of sequential runs, the energy saving percentage depends of the waiting time gap between executions.

1. Introduction

In the last few years, the usage of Graphics Processing Units (GPU) as accelerators for computing have become an important technology for High-Performance Computing (HPC).

A recent survey [1] of the top 500 supercomputers shows that eight percent of them are heterogeneous systems based on GPU cards used for General Purpose (GPGPUs). Each GPU consists of thousands of highly specialized processing cores, during computation, the CPUs offloads some compute-intensive, highly parallelizable code segments to these GPUs for execution acceleration.

The pioneer GPUs were developed for computer graphics that require high computational throughput capability, but actually they also support the execution of general purpose applications. The Compute Unified Device Architecture (CUDA) from Nvidia have been developed to support such applications on GPUs [2]. With the CUDA development, programming GPUs becomes simpler, being no longer necessary to use libraries like OpenGL.

However, the electricity costs of operating HPC systems and data centers have increased by 36 percent in the period of 2005 to 2010 (from seven gigawatts to almost ten gigawatts total in the United States), according to a 2011 report on recent electricity growth within data centers [3]. This growth was actually dampened by the global recession, it had been forecasted by the Environmental Protection Agency of the United States to double in 2016 as its Department of Energy enters the *exascale* era of supercomputing [4].

The emerging green-thinking that claims for HPC systems and programs more energetic efficiently is nowadays more present in HPC laboratories. Unfortunately, even with those efforts aiming green computing, a large part of research about the applications' energy consumption is focused mostly on application performance. To achieve the exascale era, reducing the power consumption, saving more energy and, as a consequence saving money, we need to look at the application ***post-execution*** behavior too. When an application finishes, what happens after that point is also important, as it can be wasting valuable joules.

There is a lot of effort to save energy during the application execution, some techniques are very helpful and can be adapted to reduce the consumption where is it possible yet. Improving energy efficiency is a continuous challenge in HPC because of the constant need for performance despite the constraints in the power budget and economic cost. As a result, it is necessary to invest in power-aware technologies on GPU-based heterogeneous systems.

The main motivation of this work is the fact that when a GPU application ends, its power consumption does not return as fast as possible to its lowest power draw. Our contribution is a tested mechanism to reduce the time that a GPU takes to return to its idle state, saving energy and money in the process without impacting application performance.

2. Related Work

The power consumption for GPUs have been studied using experimental methods where power is measured directly on the whole system, also using energy models or energy optimization techniques [5, 6, 7, 8].

The Dynamic Voltage Frequency Scale (DVFS) is a technique used to decrease the frequency-voltage dynamically, allowing a lower power draw. Ge et al. [9] developed a mechanism using DVFS for GPUs. However, they found an undesirable side effect of this technique, it causes a potential loss in performance when they lower the clock frequency by creating delays in the application execution. They conclude that there is a careful balance that must be struck between lowered power draw and loss of performance in order to achieve a reduced energy consumption.

The work of Lee et al. [10], although includes CPU usage, shows how to improve the throughput of power-constrained GPUs. They optimized the number of operating cores, the voltages and frequencies of caches, cores and on-chip interconnects depending on applications' characteristics. Using their method they analyzed the impact of the number of operating cores and their voltage and frequency on GPU improving the throughput nearly 20% on average.

The green-thinking brings the economic aspect related to the energy costs. To help reducing these costs, projects like Green Queue were developed. Green Queue is based on a series of application-aware Dynamic Voltage-Frequency Scaling (DVFS) techniques for use within MPI applications [11]. This scalable scheduler was deployed at the San Diego Supercomputer Center. Results of using this technique on their machines allow them to save as much as 31.7% of the operational energy bill.

In order to reduce energy waste, either for economic reasons or to reach the exascale era, techniques like frequency scaling on GPU cores and memory and workload division between CPU and GPU are used. Frameworks to control this scaling have been developed and GPGPU's are analyzed for HPC purposes. Benchmarks, algebra libraries or actual applications have been used to validate the new ideas and concepts of Green Computing [12, 13, 14]. Recently, high-end GPUs that support frequency scaling and power monitoring, such as Nvidia's K20, entered the market. The benefits of GPU frequency scaling appear like an effective approach to save system power. In general, scaling down the GPU core voltage can significantly save energy when working at appropriate core frequencies. Scaling memory frequencies can also save energy for some applications. However, it is not easy to find optimal setting for GPU frequency scaling [15].

3. Resources

The newer GPUs are not concerned only with performance, they also have several hardware counters and information about the power draw and usage. Those metrics and data can be retrieved using appropriate libraries and methods without generating a significant energy waste.

In this section, we discuss how these metrics were obtained by a developed application. Besides, we present some technical background on the Nvidia Tesla K20 and the benchmarks and applications used to evaluate our methodology.

3.1. Nvidia Tesla K20

The Nvidia Tesla K20 is a GPU card based on the Kepler architecture that comprises more than seven billion transistors adding new features focused on computing performance. The Kepler family was designed especially to attract the the HPC market. This GPU card has 2496 processor cores working at 706MHz and 5GB of memory operating at 2.6GHz.

This architecture was also designed to improve power efficiency, achieving a better performance per watt versus the previous Nvidia GPU architecture. The execution units run at a high clock rate, allowing the chip to achieve higher throughput with few copies of the execution units [16]. A trade-off of GPU cards is that their power consumption is usually higher than a CPU. The thermal design power for the K20 is 225W, while its idle power is 14W.

3.2. Nvidia NVML

Nvidia offers the Nvidia Management Library (NVML), a C-based API that allows the monitoring and managing of states in a Nvidia GPU device [17]. This library allows low-level access to the features on the GPU card, sensors data and accounting information. Also, NVML allows administrators to query the GPU device state and, with the appropriate privileges, it is possible to modify some of these GPU device states. NVML is compatible with some Tesla and Fermi devices and it has limited support on older Nvidia GPUs.

3.3. Benchmarks

To obtain the post-execution power draw behavior of applications on the K20 GPU, we used a set of applications from two benchmarks commonly used in GPU, MAGMA and Rodinia.

3.3.1. MAGMA

MAGMA is a framework to develop linear algebra applications for hybrid and heterogeneous architectures [18]. Hybrid many core and GPU systems can enable applications to fully exploit the power that each of the hybrid components offers.

MAGMA has a testing suite which comprises single and double precision test for real and complex matrix operations. It is possible to set different input sizes for each instance, that allows a wide range of processing loads and consequently a different power consumption behavior for each test [19, 20].

We used a subset of the available benchmarks of MAGMA. The tests executed with MAGMA used single and double precision, in real and complex domains, performing QR, LQ and triangular factorization operations. The matrix sizes used were 512, 1024, 2048, 4096.

3.3.2. Rodinia

Rodinia is a set of free and open benchmarks. Its applications are designed for heterogeneous computing infrastructures targeting both GPUs and multicore CPUs [21].

The Rodinia benchmarks span a range of parallelism and compute patterns, it provides applications from Linear Algebra problems to Medical Images processing. To perform the tests we used all Rodinia tests that do not require the GPU Texture Area. Each application or kernel represents a different type of behavior called dwarf. A Dwarf is a terminology proposed in Berkeley as the Dwarf taxonomy and was used to classify parallel patterns [22].

4. GPUtool

In order to gather GPU metrics, it is possible to use profilers like the Nvidia Visual Profiler [23]. However, this profiler is unable to obtain data during the post-execution interval.

To overcome this problem, we developed an application (GPUtool) to retrieve and analyze this data even after the end of the execution. GPUtool is a C-based profiling-tool running in the background using the NVML libraries.

This application retrieves the necessary metrics to analyze the energy consumption in *post-execution* from the Tesla K20 counters, without increasing significantly the power draw while running in parallel with the analyzed application.

This tool is executed in CPU collecting relevant values stored in the GPU counters. The collected metrics are, among others, execution time, power draw, GPU memory and SMX-core frequency, fan-speed and temperature. To avoid getting erroneous data about the analyzed application, it was checked that only one application was running on the GPU.

To minimize computing overhead, which could cause some power draw in the GPU, the access time to retrieve the metrics was defined to be slower than 15ms. It was confirmed by us that a faster measuring frequency increased the power draw in the card.

When the GPUtool is used only to collect data, the frequencies are controlled by the GPU included mechanisms at all time. On the other hand, when the tool is responsible for controlling the frequency at post-execution, it sets both frequencies to the lowest possible values right after the GPU application ends.

The main reason to develop a profiling application versus using one already available, was based on the lack of a tool that permits collecting data after the application finished. Available profiling tools did not work without an application context, that means that while there is no application running on the GPU it is not possible to profile and obtain some valuable metrics from the card.

5. Mechanism to reduce energy waste in Post-Execution

When a GPU application finishes, it is expected that the power draw in the GPU card return to its idle state if no other application starts. The idle state is characterized by a low power draw and both memory and core frequencies standing at their lowest values. However in some GPU devices this decrease in the power draw is not instantaneous, taking some seconds to start and some other time before reaching that state to avoid unnecessary energy consumption.

The mechanism to reduce this return to idle time and its subsequently energy waste is to force a frequency scaling when the post-execution phase starts. In this way, both memory and core frequencies are set to their lowest values, resulting in a fast reduction in energy consumption. By applying the frequency scaling at the start of the post-execution phase, both core and memory frequencies are set to the lower states, resulting in a quick decrease of power draw. This process reduces the energy wasting in this non-computing state.

The applications we used to validate our mechanism were executed on the GPU without frequency limitations, allowing the GPU energy management to set the adequate frequency by default. The highest values of the frequencies are 2600MHz for the memory and 706MHz for the GPU cores. The lowest values they both could be set are 324MHz.

This mechanism can be analyzed in two scenarios with a difference at what happens after the post-execution phase starts. The first scenario is when there are no more applications waiting to be executed and the second one when there are consecutive executions of GPU applications. In both scenarios, it was analyzed the post-execution behavior while implementing the proposed mechanism. In the next subsections both scenarios will be detailed.

5.1. Single run scenario

This scenario consists of the execution of a GPU application and its posterior post-execution phase until it reaches the idle state. In this scenario no other application starts during the post-execution phase. This setting captures the behavior of the post-execution energy consumption without the effects of more executions of GPU applications.

Figure 1 shows a timeline with each one of the phases of a GPU application run. In this scenario GPUtool activates the energy saving mechanism at the only post-execution phase.

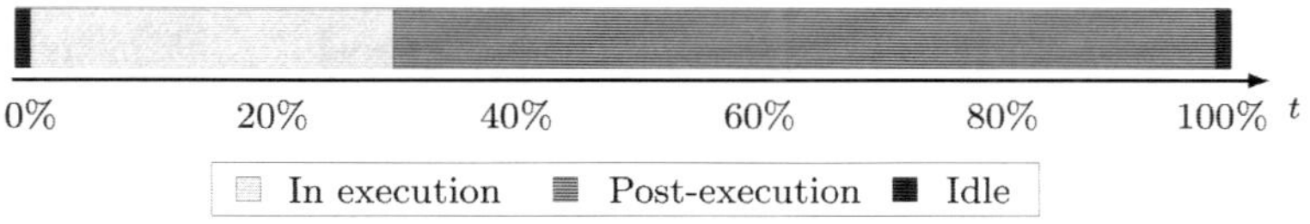

Figure 1. Phases in a single run scenario.

5.2. Multiple run scenario

This scenario consists in the consecutive execution of one GPU application, starting each execution during the post-execution phase of the previous one. This creates a behavior similar to an scheduler, with several applications running in the GPU.

Figure 2 depicts a detailed timeline for each of the phases taken into account in the multiple run scenario. The methodology to save energy is applied only at the post-execution intervals, i.e., between two consecutive executions and at the last post-execution interval until reaching the idle state.

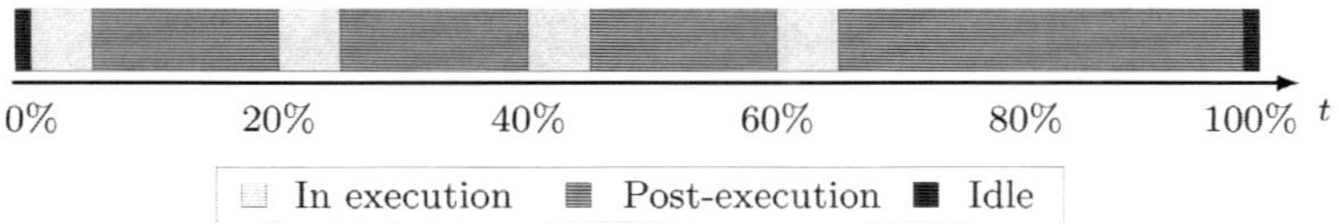

Figure 2. Phases in a multiple run scenario.

The multiple run scenario was considered to illustrate that the proposed mechanism does not generate a significant performance loss in posterior executions. Because GPUTool does not know beforehand how many applications will be executed or the duration of them, the tool needs to be launched before the application execution starts to act when it is necessary.

Since this tool does not produce a significant impact on application performance along the application execution, it is possible to save energy at every moment that there are no applications running on the GPU without compromising performance.

6. Results

All tests were executed on two different machines using four different devices, two Nvidia Tesla K20m and two Nvidia Tesla K20c. In order to discard manufacturing problems with the cards and to compare the results between them. All the results presented here are the arithmetic average of multiple executions.

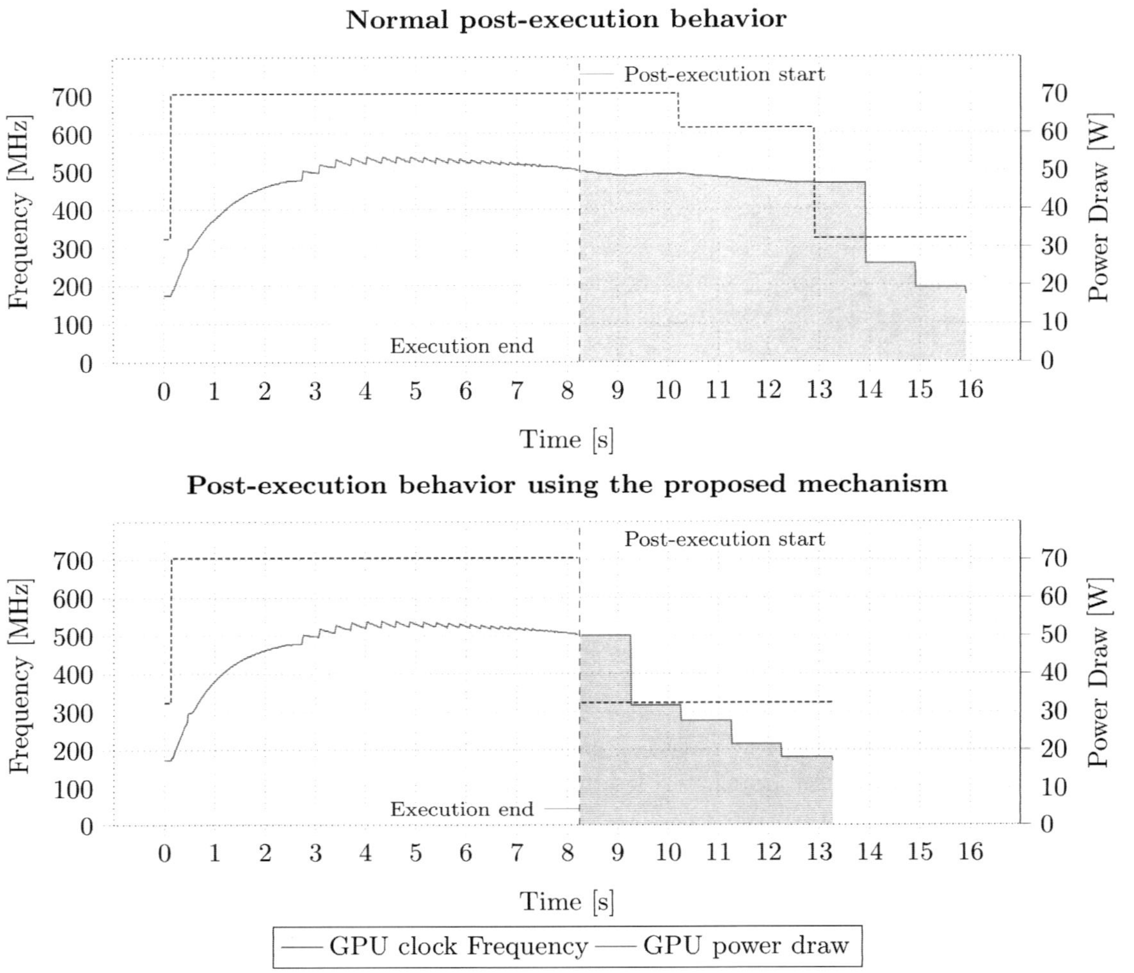

Figure 3. Energy consumption of CGEQRF with input size 4096 in a single run scenario. Note that in the normal post-execution behavior the idle state is reached near 16 seconds, on the other hand, using the proposed mechanism this state is achieved in a shorter time with a lower power draw, minimizing the energy waste. The dark area represents the energy consumption.

A comparison of an execution of the benchmark CGQRF and its power draw with and without our mechanism is depicted in Figure 3. In the Figure, the power draw is the same in both in-execution phases because the mechanism is activated only after the start of the post-execution phase. The vertical dashed line indicates where this phase starts. The area under the GPU power draw curve represents the energy consumption. The dark area allows a visual comparison of the post-execution savings. The power draw stays near its last value before the application ended, returning to idle after some seconds.

Without the proposed mechanism, the GPU card persist in a high frequency during non-computing periods and this high frequency consumes unnecessary power for GPU card. With the mechanism active, both frequencies are returned to their lowest values and the power draw is reduced quickly. Savings in the post-execution of this benchmark were 52.96 percent. As shown in the Figure, the execution time is not altered by the mechanism, allowing energy savings in the post-execution phase of a single run without affecting the application in-execution behavior.

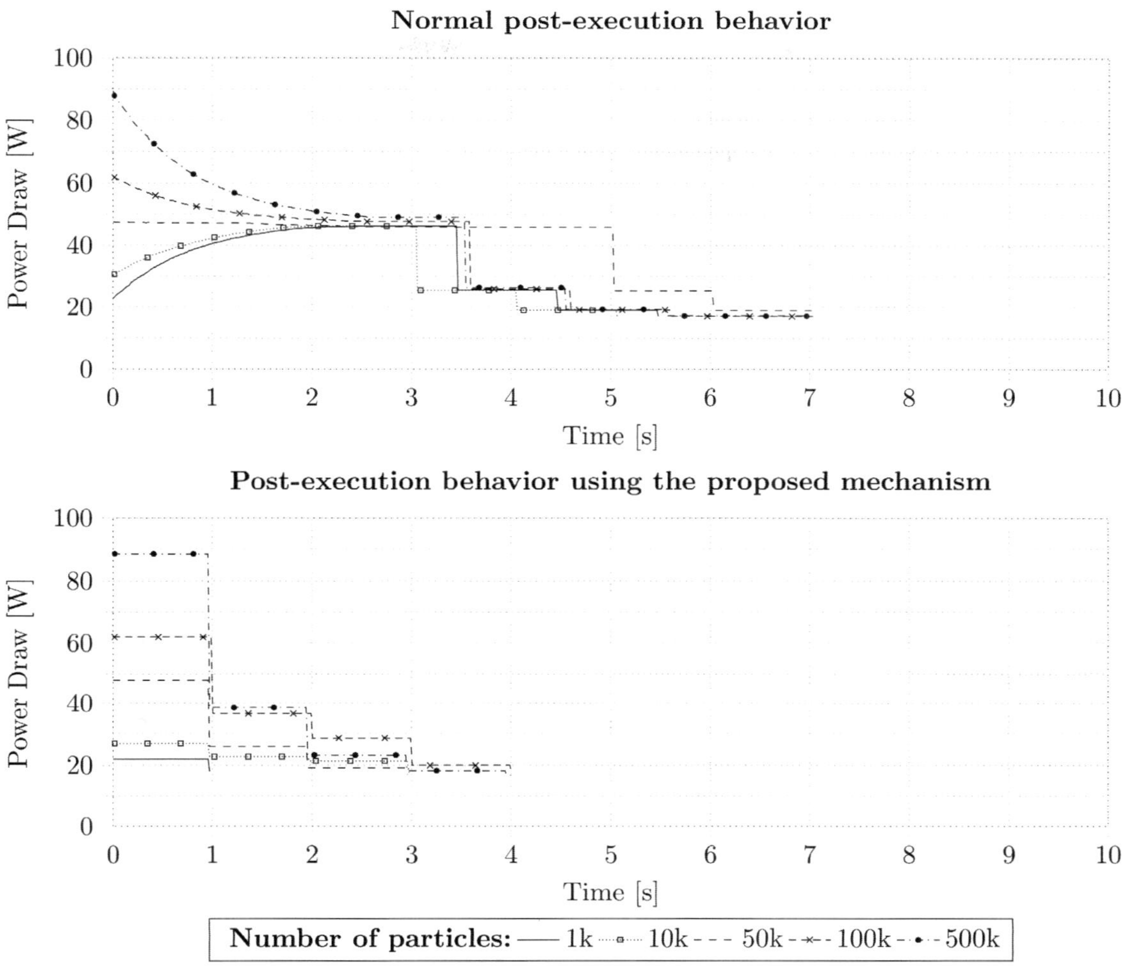

Figure 4. Post-execution energy consumption using different input data size to achieve different power draw behaviors for Particle Filter benchmark. Time zero indicates the start of the post-execution phase.

A more detailed view of the post-execution behavior is showed in Figure 4. The Figure depicts only the post-execution phase of the Particle Filter benchmark using different number of particles as input size. As the figure shows, the time-to-idle period is different based on the input size, it seems to be a high correlation between the power draw at the end of an application execution and the post-execution phase length.

The post execution savings of Rodinia's Particle Filter benchmark, with and without the usage of the mechanism in the same order as the input sizes of Figure 4 are 88.55%, 59.91%, 67.03%, 35.12%, 33.33%. As illustrated in Figure the clock frequency behavior right after the end of the execution impacts on power draw. That means that the time that power draw takes to return to idle and the post-execution savings depends on the power draw at the very end of the execution.

Using the MAGMA benchmarks in all of them we obtained savings during its post-execution phase. Figure 5 shows the normalized power savings of the MAGMA tests. As depicted in the Figure, in some benchmarks our mechanism is able to achieve energy savings up to 73 percent.

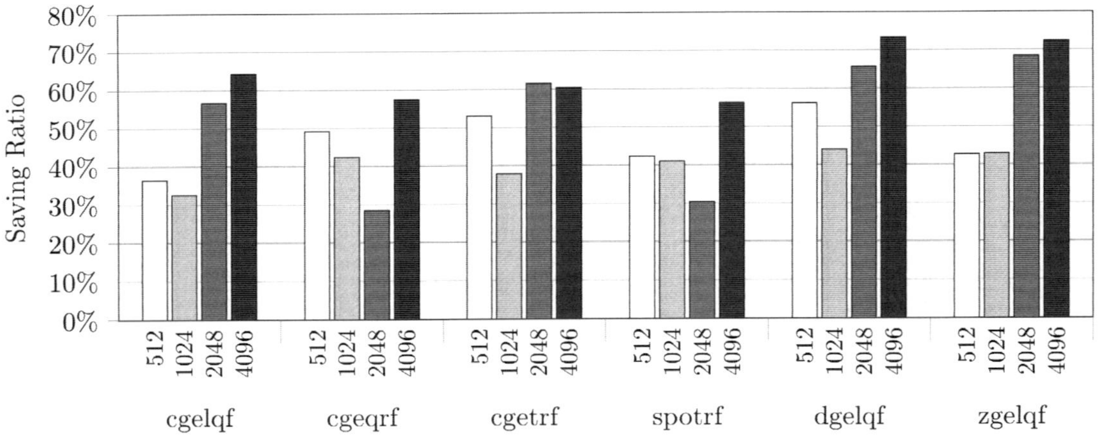

Figure 5. MAGMA benchmarks. Average saving ratio of each benchmark using different input data sizes.

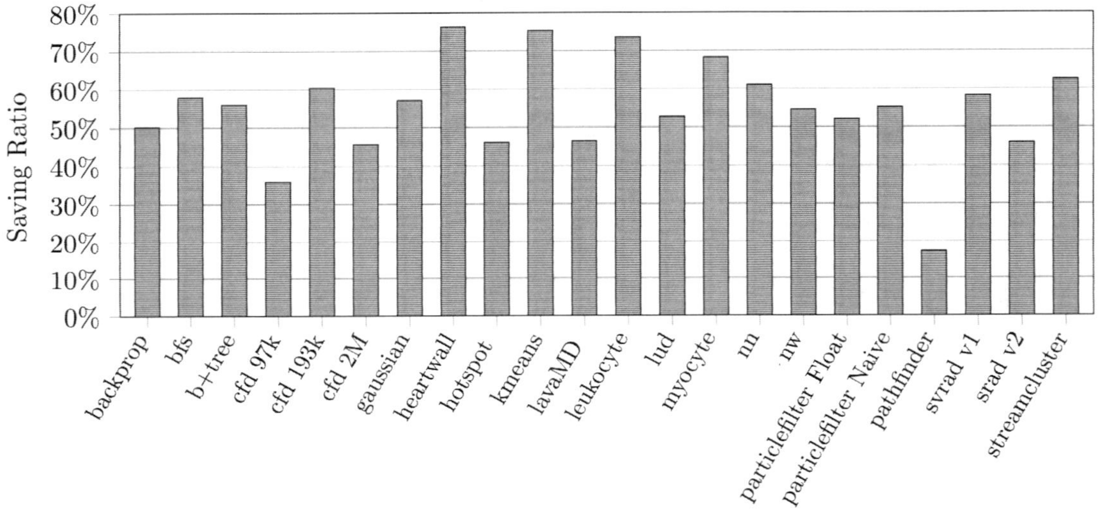

Figure 6. Rodinia benchmarks. Average saving ratio of each benchmark using the default input size.

In relation to the minimum savings, the MAGMA benchmark with the lowest value reached 28.45 percent of energy savings, even this value could be considered as a good result.

The post-execution behavior of this benchmark also allowed us to confirm that applications with higher power draw at the end of execution achieved bigger savings. In all of the experiments using MAGMA benchmarks our mechanism reduced the power consumption during the post-execution phase.

Regarding the Rodinia benchmarks we obtained the following results. Using the Heart Wall Tracking benchmark with its default parameters, our mechanism reduced 65.59 percent the time to idle, saving more than 245J. In the Back Propagation benchmark were achieved energy savings of 50.22 percent representing 94.29J. Saving ratios of the Rodinia benchmarks are depicted in Figure 6. As in the previous set of benchmarks, in all of the cases we were able to reduce power consumption during the post-execution phase.

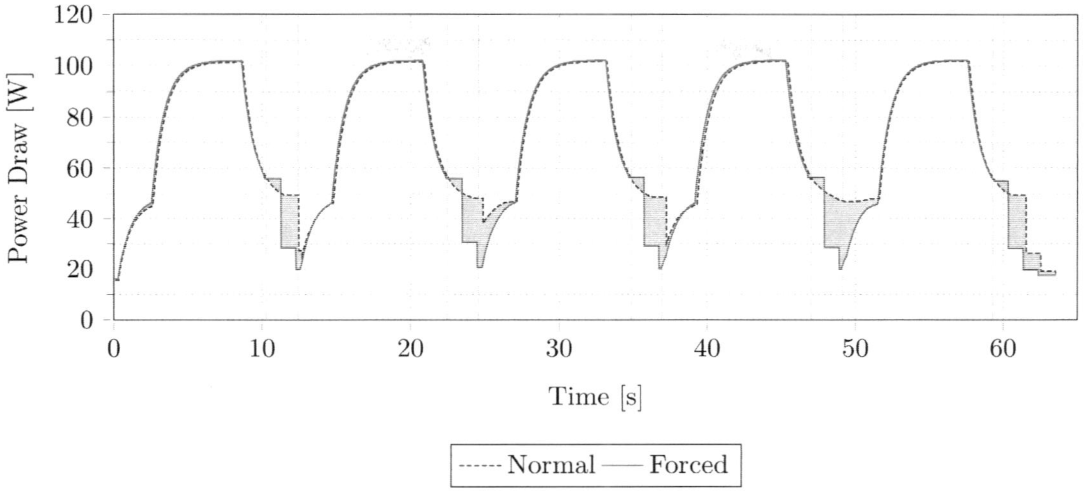

Figure 7. Energy consumption on five executions of the CFD benchmark with 2 million elements. Two seconds of post-executions phase between executions. The final post-execution phase goes until idle state. The dark area between plot lines represents the total energy savings.

As proposed for the second scenario, in order to reduce the power consumption between two runs we limited the waiting time between the consecutive executions of the same GPU application. This scenario is shown in Figure 7 using CFD, one of the Rodinia benchmarks, with an interval of two seconds between five executions. In this benchmark we achieved savings of 118J in the post-execution phase, almost 21 percent. As depicted in the Figure, application performance was not impacted by using our mechanism between the executions. Because the mechanism is activated as soon as the GPU application finishes, if the post-execution runs for a significant amount of time it will allow greater energy savings. However, it is not realistic to wait a lot of time between executions when running short-time GPU applications sequentially.

The results obtained shows that is possible to obtain savings in the two scenarios proposed in our methodology, without impacting performance.

7. Conclusions

In this work we conducted a measurement study of the power consumption savings by analyzing the behavior of the GPU card after a running application finished. We used some benchmarks and variations in their input sizes to generate different post-execution behavior. We demonstrated that power waste exists in GPU cards and that this could be mitigated using frequency scaling. While frequency scaling already is a common strategy to save energy during the application' execution, there are no previous works performed in post-execution energy savings using this approach.

For short running applications, scaling down the GPU and Memory frequency can significantly save energy during the post-execution phase. Using this approach, savings up to 73 percent could be achieved. This mechanism could be implemented directly on the GPU power manager, optimizing the resources usage for every GPU application.

By aiming to reduce the energy consumption of configurable GPUs, our approach can be a starting point to claim the attention for an overlooked area of energy waste, the post-execution consumption. These results could be useful in the road to the exascale era where there are efforts to save every watt as possible for computation.

8. Future Work

In the future, an analysis of the consumption behavior when several applications are executed in parallel could be performed. With this approach it could be possible to reduce power consumption in the execution gaps, that is when there are no kernels running in the device, extending our mechanism to in-execution savings. Also, it may be analyzed the behavior of the applications that require more than one GPU device.

Acknowledgments

The authors would like to thank all the members of the Parallel and Distributed Processing Group (GPPD) at Federal University of Rio Grande do Sul (UFRGS), their help, expertise and advice were of great value.

This research have been partially supported by CNPq under Grant 132123/2013-4, by FAPERGS under grant ME 14/2012 and by the UFRGS Informatics Institute.

References

[1] Top500.org, "China's Tianhe-2 Supercomputer Maintains Top Spot on 42nd TOP500 List," http://www.top500.org/blog/lists/2013/11/press-release/, 2013.

[2] Nickolls, John and Buck, Ian and Garland, Michael and Skadron, Kevin, "Scalable Parallel Programming with CUDA," *Queue*, vol. 6, no. 2, pp. 40–53, Mar. 2008. [Online]. Available: http://doi.acm.org/10.1145/1365490.1365500

[3] Koomey, Jonathan, "Growth in data center electricity use 2005 to 2010," *Oakland, CA: Analytics Press. August*, vol. 1, p. 2010, 2011.

[4] Brown, Richard and others, "Report to Congress on Server and Data Center Energy Efficiency Public Law 109-431," *Lawrence Berkeley National Laboratory*, vol. 109, p. 431, 2007.

[5] Chen, Jianmin and Li, Bin and Zhang, Ying and Peng, Lu and Peir, Jih-kwon, "Statistical GPU power analysis using tree-based methods," in *Green Computing Conference and Workshops (IGCC), 2011 International.* IEEE, 2011, pp. 1–6.

[6] Kasichayanula, Kiran and Terpstra, Dan and Luszczek, Piotr and Tomov, Stan and Moore, Shirley and Peterson, Gregory D, "Power aware computing on GPUs," in *Application Accelerators in High Performance Computing (SAAHPC), 2012 Symposium on.* IEEE, 2012, pp. 64–73.

[7] Ukidave, Yash and Kaeli, David R, "Analyzing Optimization Techniques for Power Efficiency on Heterogeneous Platforms," in *Parallel and Distributed Processing Symposium Workshops & PhD Forum (IPDPSW), 2013 IEEE 27th International.* IEEE, 2013, pp. 1040–1049.

[8] Ukidave, Y. and Ziabari, AK. and Mistry, P. and Schirner, G. and Kaeli, D., "Quantifying the energy efficiency of FFT on heterogeneous platforms," in *Performance Analysis of Systems and Software (ISPASS), 2013 IEEE International Symposium on*, April 2013, pp. 235–244.

[9] Ge, Rong and Vogt, Ryan and Majumder, Jahangir and Alam, Arif and Burtscher, Martin and Zong, Ziliang, "Effects of dynamic voltage and frequency scaling on a k20 gpu," in *2nd International Workshop on Power-aware Algorithms, Systems, and Architectures*, 2013.

[10] Lee, Jungseob and Sathisha, Vijay and Schulte, Michael and Compton, Katherine and Kim, Nam Sung, "Improving throughput of power-constrained GPUs using dynamic voltage/frequency and core scaling," in *Parallel Architectures and Compilation Techniques (PACT), 2011 International Conference on.* IEEE, 2011, pp. 111–120.

[11] Ananta Tiwari and Michael Laurenzano and Joshua Peraza and Laura Carrington and Allan Snavely, "Green Queue: Customized Large-Scale Clock Frequency Scaling," in *Cloud and Green Computing (CGC) , 1-3 November , 2012*, IEEE. Xiangtan, Hunan, China: IEEE, 2012. [Online]. Available: http://users.sdsc.edu/~lcarring/Papers/2012_CGC.pdf

[12] Padoin, Edson Luiz and Pilla, Laércio Lima and Boito, Francieli Zanon and Kassick, Rodrigo Virote and Velho, Pedro and Navaux, Philippe OA, "Evaluating application performance and energy consumption on hybrid CPU+ GPU architecture," *Cluster Computing*, pp. 1–15, 2012.

[13] Ma, Kai and Li, Xue and Chen, Wei and Zhang, Chi and Wang, Xiaorui, "Greengpu: A holistic approach to energy efficiency in gpu-cpu heterogeneous architectures," in *Parallel Processing (ICPP), 2012 41st International Conference on.* IEEE, 2012, pp. 48–57.

[14] Huang, Song and Xiao, Shucai and Feng, Wu-chun, "On the energy efficiency of graphics processing units for scientific computing," in *Parallel & Distributed Processing, 2009. IPDPS 2009. IEEE International Symposium on.* IEEE, 2009, pp. 1–8.

[15] Mei, Xinxin and Yung, Ling Sing and Zhao, Kaiyong and Chu, Xiaowen, "A measurement study of GPU DVFS on energy conservation," in *Proceedings of the Workshop on Power-Aware Computing and Systems.* ACM, 2013, p. 10.

[16] Nvidia, "Whitepaper Nvidia's Next Generation CUDA Compute Architecture: Kepler GK110," http://www.nvidia.com/content/PDF/kepler/Nvidia-Kepler-GK110-Architecture-Whitepaper.pdf, 2012.

[17] Nvidia, "Nvidia Management Library (NVML)," https://developer.nvidia.com/nvidia-management-library-nvml, 2013.

[18] Haidar, A. and Tomov, S. and Yamazaki, I. and Solca, R. and Schulthess, T. and Dong, T. and Dongarra, J., "MAGMA: A Breakthrough in Solvers for Eigenvalue Problems," http://icl.utk.edu/magma/, 2008.

[19] Agullo, Emmanuel and Demmel, Jim and Dongarra, Jack and Hadri, Bilel and Kurzak, Jakub and Langou, Julien and Ltaief, Hatem and Luszczek, Piotr and Tomov, Stanimire, "Numerical linear algebra on emerging architectures: The PLASMA and MAGMA projects," in *Journal of Physics: Conference Series*, vol. 180. IOP Publishing, 2009, p. 012037.

[20] Innovative Computing Laboratory - UTK, "Matrix Algebra on GPU and Multicore Architectures," http://icl.utk.edu/magma/, 2008.

[21] Che, Shuai and Sheaffer, Jeremy W and Boyer, Michael and Szafaryn, Lukasz G and Wang, Liang and Skadron, Kevin, "A characterization of the Rodinia benchmark suite with comparison to contemporary CMP workloads," in *Workload Characterization (IISWC), 2010 IEEE International Symposium on.* IEEE, 2010, pp. 1–11.

[22] Che, Shuai and Boyer, Michael and Meng, Jiayuan and Tarjan, David and Sheaffer, Jeremy W and Lee, Sang-Ha and Skadron, Kevin, "Rodinia: A benchmark suite for heterogeneous computing," in *Workload Characterization, 2009. IISWC 2009. IEEE International Symposium on.* IEEE, 2009, pp. 44–54.

[23] Nvidia, "Nvidia Visual Profiler," http://developer.nvidia.com/nvidia-visual-profiler, 2013.

Code optimization by using GPU applied to a Dataflow numerical simulation model

Luiz E S Evangelista [1], Alvaro L Fazenda, Vincius V de Melo

Science and Technology Institute, Federal University of Sao Paulo (UNIFESP)
Avenida Cesare Mansueto Giulio Lattes, 1201 - 12247-014 - São José dos Campos - SP - Brazil

E-mail: Luiz_evangelista@hotmail.com, alvaro.fazenda@unifesp.br and vinicius.melo@unifesp.br

Abstract. This work aims to study techniques for parallel computing using GPU (Graphics Processing Unit) in order to optimize the performance of a fragment of computational code, implemented as a Dataflow system, which is part of a meteorological numerical model responsible for calculating the advection transportation phenomena. The possible algorithm limitations for GPU efficiency will also be addressed through an extensive code instrumentation. Considering the difficulties found on the original algorithm which implies a GPU accelerated code dealing with flow dependencies and coarse grain parallelism, the performance gain with GPU may be considered fair.

1. Introduction

The data processed in numerical simulation can reach big volumes nowadays. Thus parallel programming is widely used, due to the large amount of time to get the expected results. Like other areas in Computer Science, such programming paradigm evolved and currently is possible to use external devices like GPUs as arithmetic co-processors in a fine-grain parallelism model [11].

A simple way of programming a GPU consists in using a programming language that follows the OpenACC standard [16]. Such language provides a set of directives to be applied to the original source-code, allowing the programmer to define a specific part of the code to be accelerated by an external device. The major benefit of this programming model is the lower effort to port the original code, created to be used with traditional CPUs, when compared to the popular CUDA language [11].

This paper aims to optimize the performance of a source-code developed following a data-flow approach, used to numerically simulate the advection transportation phenomena through a code ported to GPU using the OpenAcc standard. This numerical model is used as part of a weather forecast and climatic model. There are several successful cases of numerical simulation accelerated by GPUs, but the existing data-flow approach in the code represents a significant difficulty to achieve satisfactory efficiency. The paper shows a detailed performance analysis for the developed code aiming to identify major performance bounds, differences in algorithm performance applied to data arranged in different coordinate axes and scalability related to domain size and meteorological fields increments. The code robustness are evaluated through powerful accelerators and higher domains. The performance behavior under different kind of synthetic data are also measured, trying to identify better and worse case performance.

[1] To whom any correspondence should be addressed.

Hartley et al [7] propose a new scheme to solve a particular problem in a hybrid system using CPUs and GPUs, through a task division method using a data-flow, in which it is possible to dynamically change the task loads in order to keep the system balanced. Boulos et. al [1] solve a signal processing problem by using some GPUs with a model based on a data-flow approach, launching concurrent jobs on GPU overlapping computation and communication between host and device. Wang et. al. [20] addresses the development of a data-flow system in GPU applied to a signal processing application. Meng et. al [13] shows a framework to generate GPU code based on a data-flow scheme, by analyzing loops in the source-code.

There are just a few numerical meteorological models fully ported to GPUs. Cumming et. al. [2] have accelerated a meteorological model called COSMO by using GPU through code refactoring. On that work, part of the model responsible for fluid dynamics uses an API created in C++ and CUDA language, while the other part of the model, responsible for physics simulation phenomena, was ported to GPU by using OpenAcc standard directives.

Michalakes and Vachharajani [14] show a part of the WRF (Weather Research and Forecasting) model used in physics simulation phenomena ported to GPUs in CUDA language. This portion of code reached a speedup of about 20, with 7.4 Gflops.

Wang et. al. [20] present an approach for porting three different subroutines of GRAPES model (Global/Regional Simulation Prediction System) to GPU by using CUDA Fortran language. They highlight the three following key factors to improve performance on GPUs: reducing data transfer time between host and device, reducing the amount of memory access, and a thread control flow avoidance mechanism.

Govett et. al. [6] describe the module responsible for computational fluid dynamics in the NIM model (Non-hydrostatic Icosahedral Model), adapted to be used in GPUs. The authors also wrote a compiler to convert FORTRAN into CUDA code called F2C-ACC (FORTRAN-to-CUDA ACCelerator), which works using directives. The ported module runs 34 times faster on a single GPU than on the CPU used in the comparison.

In Shi [17] Master's thesis the performance analysis and optimization to GPUs was applied to serial solvers used to evaluate several numerical procedures in the Met Office Unified Model. The speedup by using CUDA language reach 10 times.

In Haines [8] Master's thesis, several physics schemes from the WRF model have been ported to GPU by applying OpenAcc directives on the original code. According to the author, the preliminary results show a modest 1.2x speed-up by simply introducing approximately 20 lines of OpenAcc directives.

Other work focus just in optimization by GPUs applied to advection transportation phenomena. Some of them could be also applied to weather forecast models or just integrate a computational fluid dynamics software. Souza et al [18] describe a parallel finite element implementation for the 2D time-dependent advection-diffusion problem ported to GPU, and the relationship between performance and energy consumption when compared to a parallel version working on a CPU cluster. White and Dongarra [22] have written a code to solve the advection problem that overlaps processing and communication on GPUs and describe results for FORTRAN implementations using various combinations of MPI, OpenMP, and CUDA.

All related work depicted here differs from this current paper by the numerical method adopted. The Walcek numerical method [19] to solve the advection transportation phenomena imply in data-streams, which was mapped to a dataflow scheme by Fonseca et al [4] as a computational strategy to allow concurrency. In this work this problem was ported and optimized to work with GPUs.

In the following sections the readers will find this work organized as follows: Chapter 2 describes the monotonic advection algorithm, Chapter 3 shows the Data-Flow system applied to the original advection algorithm, Chapter 4 presents the results, and finally in Chapter 5 the discussion and conclusion are postulated.

2. Algorithm to evaluate the Advection phenomena

The numerical advection model is a fundamental part for any fluid dynamics model. In the meteorology context the advection phenomena is usually responsible for the transport of scalar properties (e.g., mass density and mass mixing ratio) by the wind flow [5].

A new numerical method to evaluate the advection phenomena in a particular weather forecast model was proposed by Freitas et al. [5] and it will be referred in this work as monotonic advection. This new scheme is applied to the Coupled Chemistry-Aerosol-Tracer Transport model to the Brazilian developments on the Regional Atmospheric Modeling System (CCATT-BRAMS) [12], and contribute to more accurate results in numerical simulations over tropical and sub-tropical regions of South America. Those better results can improve the services available from CPTEC/INPE (Center for Weather Forecasting and Climate Research/National Institute for Space Research) in Brazil, where this model runs daily to simulate atmosphere pollution due gases emissions in urban centers and forest burns.

The first computational implementation of monotonic advection scheme applied in CCATT-BRAMS model has used a serial code without any parallelism [4], due to several dependencies in the original numerical algorithm. In Algorithm 1 it is possible to see part of this original code including two main loops, which iterate over an array in both positive and negative directions representing the discretized domain of the problem. The first loop (Algorithm 1) evaluates the advection transportation over adjacent cells with positive value for the wind velocity ($u_i > 0$) in a particular coordinate axis. The second loop does the same for the adjacent cells where the wind velocity is negative ($u_i < 0$). However, in the first loop there is a flow dependency related to instructions: $Qn_i = Fa(Flux_i, ...)$ and $Flux_i = Fb(Flux_{i-1}, Qn_i, ...)$. In the second loop, a similar dependency appears on instructions: $Qn_i = Fa(Flux_i, ...)$ and $Flux_{i-1} = Fb(Flux_i, Qn_i, ...)$, where Fa and Fb represent arithmetical procedures.

Algorithm 1 Pseudocode to evaluate the advection in an array [4, 19]

```
Flux evaluation from right array boundary
for i=2,n-1 do
    if u_i < 0 then
        Cycle
    end if
    if vel_{i-1} < 0 then
        flux_i = F(u_i, ...)
    else
        Qn_i = Fa(Flux_i, ...)
        Flux_i = Fb(Flux_{i-1}, Qn_i, ...)
    end if
end for
Flux evaluation from left array boundary
for i=n-1,2,-1 do
    if u_{i-1} ≥ 0 then
        if u_i < 0 then
            Qn_i = F(flux_i, Flux_{i-1}, ...)
        end if
    else
        Qn_i = Fa(Flux_i, ...)
        Flux_{i-1} = Fb(Flux_i, Qn_i, ...)
    end if
end for
```

3. Data-flow scheme applied to advection

To deal with flow dependencies in the original form of monotonic advection (Figure ??) implemented a Data-flow approach which detects and evaluates concurrent streams (sequence of adjacent cells with same wind direction) over the computational discrete domain. The first step to create the data-flow structure consists in identifying cells based on their wind direction neighborhood, and classifying each one like shown in Table 1.

The Figure 1 show an example for an array representing wind velocity direction, where each cell was classified according to the labels depicted in Table 1. The next step consists in joining cells with same wind direction as streams. The evaluation order of the labeled cells respecting the flow dependencies could be represented by a directed acyclic graph (DAG) [4], as shown in Figure 2.

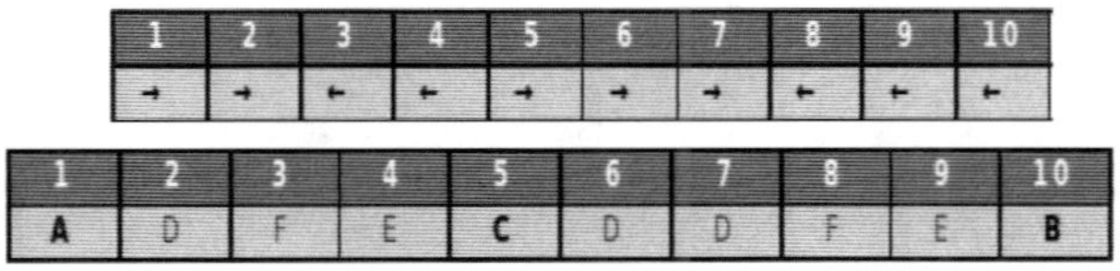

Figure 1. Classified wind array [4].

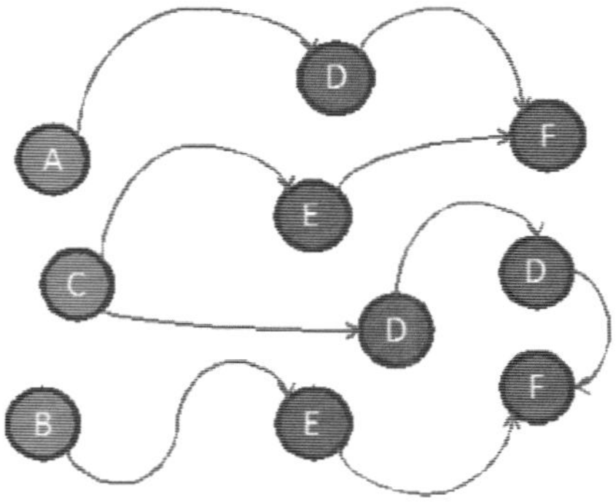

Figure 2. DAG representing flux dependencies among cells [4].

Table 1. Cell labels

$Vel(i-1)$	$Vel(i)$	Label	Description
X	$\rightarrow$	A	Inflow Left Border
$\leftarrow$	X	B	Inflow Right Border
$\leftarrow$	$\rightarrow$	C	Only Outflow
$\rightarrow$	$\rightarrow$	D	Positive stream
$\leftarrow$	$\leftarrow$	E	Negative stream
$\rightarrow$	$\leftarrow$	F	Only Inflow

The DAG in Figure 2 can be used in order to aggregate cells in streams, and can be represented by a Heap. In the Figures 1 and 2 it is possible to identify three concurrent independent streams and two dependent cells, as depicted in Table 2.

Table 2. Independent Streams and dependent cell detected

Independent Streams	Dependent Cells
a) AD (1,2)	d) F (3)
b) ECDD (4,5,6,7)	e) F (8)
c) EB (9,10)	

The algorithm responsible to preprocess the data, identifying and classifying the streams/tasks, is $O(XYZ)$, where X, Y and Z represent the total points in each axis for three-dimensional discrete domain. The streams are inserted in a heap or array structure following the order: data in X axis in head, data in Y axis in the middle and data in Z axis in tail. For each axis the concurrent streams are in head and the dependent streams are in tail, in order to respect the dataflow scheme. The elapsed time to execute the preprocessing is minor than 0.5% of total time, for the biggest case evaluated.

It is important to notice that the tasks in Table 2 require different computational effort to be done due to stream length and possible different memory arrangement. These characteristics generally represent challenges to achieve high efficiency in GPUs.

The monotonic advection source-code can be applied to any array with a specific offset between adjacent cells for each coordinate axis. A three-dimensional array representing the discretized physical domain was used in all following tests, thus the application of the same algorithm to each axis requires just the correct specification for the first element and the offset to the next cells. The computational memory allocation order is organized as ZXY to follow the same scheme in the original meteorological model coded in FORTRAN. However the same numerical procedure must be applied to N scalar meteorological fields [5]. Therefore, the final array memory arrangement is $NZXY$, which is the same computational complexity $O(NZXY)$ for the monotonic advection algorithm applied to a three-dimensional domain.

The heap proposed in the data-flow approach for monotonic advection in order to address the tasks was substituted by an array of tasks in the GPU. The array follows the same heap order, which contains all concurrent tasks at the head and the dependent tasks at the tail. It is possible to check in Algorithm 2 and 3 the original serial code in CPU and the accelerated parallel code, respectively.

Algorithm 2 Original serial code - Tasks list stored in a heap

```
while not(end(Heap)) do
    taskget_job(Heap)
    advection(task)
    update_heap(Heap)
end while
```

4. Results

In order to improve the model performance, different versions were developed (all of them written in FORTRAN-90 language). Each new release implements improvements on data structure,

Algorithm 3 Accelerated parallel code - Tasks list in an array

```
Data transfer CPU → GPU(4Darrays, Taskslist)
for (concurrent tasks) do                          ▷ GPU Kernel
    advection(tasks)
end for
for (dependent tasks) do                           ▷ GPU Kernel
    advection(tasks)
end for
Data Transfer GPU → CPU(4Darrays)
```

numerical algorithm, data management on GPU, or in OpenACC directives. All versions developed are described in this work in order to clarify the optimization processes and to inspire methods to guide future efficient GPU code porting for other similar numerical algorithms.

All data advected by the numerical method was contained in a four-dimensional array arranged as $NXYZ$, where N represents the number of scalar meteorological fields and X, Y, and Z represent the total discrete points in each coordinate axis, respectively.

The developments were performed using two different machine configurations. Most tests used an CPU *Intel Core i7-975* with accelerator *NVidia Tesla C2050* , and a PGI Accelerator v.14.1-0 compiler which supports OpenACC directives. Another machine used only for more computationally expensive tests has a 12 cores *Intel Ivy Bridge* with an *NVidia Tesla K20X* accelerator, and PGI Accelerator v.14.7 compiler. Details about the different model versions/releases used along this work are described in the following items:

(1) **Original serial code in CPU [4]:** Original serial data-flow code to be executed in a CPU and used as a performance reference.

(2) **Parallel code with OpenMP:** Adaptation of the original serial code using OpenMP directives, which allows concurrent tasks advecting streams evaluation. The source-code requires a critical section responsible for heap management and synchronization points to comply with task dependency.

(3) **Naive code with OpenACC:** First version in GPU with basic (naive) OpenACC directives, such as data transfer sections and GPU kernels over the numerical method. Unnecessary data transfer was noticed in this source-code.

(4) **GPU logical data-structure tuned code:** Code tuned for GPU blocks and threads by setting OpenACC Gang and Vector data-structure clauses, where the outer loop (responsible for tasks list iteration) was fixed to use 512 blocks and the inner loop (responsible for traversing the meteorological fields) was fixed to use 32 threads. This configuration was determined after experimental tests. It is worth emphasizing the need to inform the compiler that these loops are totally independent (independent OpenACC clause), otherwise the compiler automatically defines dependency where they do not exist.

(5) **Data transfer optimization code:** Code tuned by intent optimal definition for all data used in the numerical simulation. The data intent defines different kinds of data transfers from host to device and vice-versa (copyin, copyout, copy and create OpenACC clauses). Data transfer reduction was also implemented in this version specially applied in the interval between each axis.

(6) **Better GPU occupancy code:** In this version a previous evaluation of a small part of code was performed out of main outer loop in order to allow both a better GPU occupancy [11] and better performance. Such modification requires more memory on the system. Regarding code portability, it does not cause significant performance changes when running in traditional CPUs.

(7) **Branches decrease code:** In this version the code gets a replacement of conditional structures (Ifs) by masked arithmetic expressions based on flags. This action reduced the average of divergent paths in 39.4% in a given problem tested, with improvement in performance.
(8) **Sorted stream code:** Version developed to sort the streams by a task rank according to their size and stream type, trying to reduce divergent branches in operations within a GPU Warp [11].
(9) **Hybrid code:** Code prepared to use simultaneously multicore CPU and GPU to perform tasks.
(10) **Previous calculation code:** Version joining two phases of the code: the first phase calculates everything that is possible with independent parallel floating point operations by each grid cell, and the second phase uses the data produced by the first phase as input to evaluate the advection following the data-flow approach over streams.

The first seven versions developed show a continuous performance increase, as may be observed in the following. Unfortunately, the last three versions did not increase performance when compared to version seven due to several reasons that will be detailed in Section 4.1. Figure 3 presents the speedup for the first seven successful versions applied to a problem with the following dimensions: $22 * 38 * 336 * 336(NZXY)$. The input data was obtained from a real meteorological model simulation.

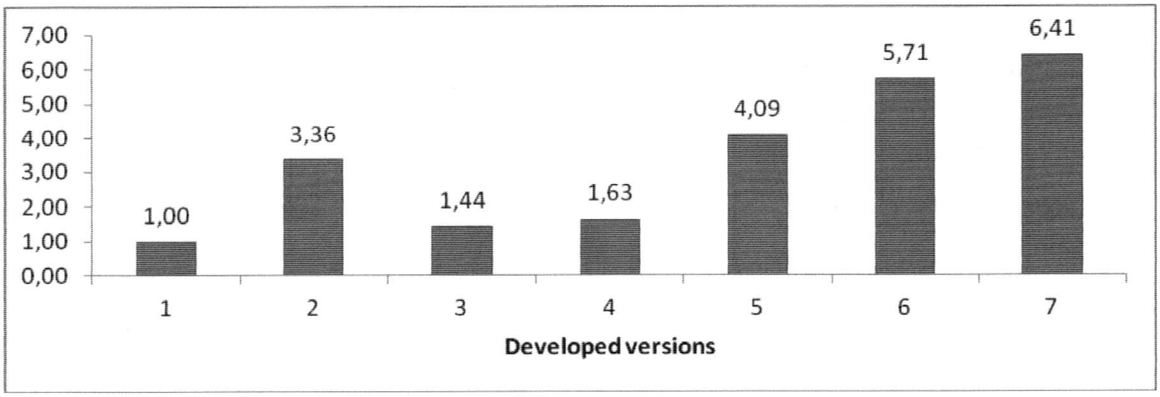

Figure 3. Speedup for the first seven stable versions

Besides the speedup, there are other evidences about the continuous performance evolution for the versions developed for GPU, like some metrics and events measured by a profiling tool called "nvprof" [3]:

- **ipc:** Instructions executed per cycle.
- **gld_efficiency:** Efficiency for load data requests in GPU global memory.
- **gst_efficiency:** Efficiency for store data requests in GPU global memory.
- **achieved_occupancy:** Real Occupancy achieved in GPU.
- **MIPS:** Millions of instructions per second.

Table 3 shows the above metrics for same seven developed GPU versions, considering a problem configured as described in the test for Figure 3. The IPC metric clearly shows successive performance increases along the versions which allow an increased number of operations executed

Journal of Physics: Conference Series **649** (2015) 012003 doi:10.1088/1742-6596/649/1/012003

per cycle. It is also possible to see the decrease in the use of branches (version seven) linked to an improvement in read access to global memory. The algorithmic modification (version six) which provides a better GPU occupancy also improves write access to the global memory. Furthermore, the Gigaflops rate shows continuous performance increase for all successive first seven versions.

Table 3. Metrics and events measured by nvprof for the first seven versions

Version	ipc	gld_efficiency	gst_efficiency	achieved_occupancy	MIPS	Gigaflops
3	0.421	44.3%	46.0%	7.1%	416896716.2	6.2
4	0.631	39.0%	23.5%	15.3%	478748992.8	7.1
5	0.641	42.7%	42.7%	15.3%	691176079.5	17.8
6	1.061	42.6%	42.6%	30.3%	835767651.2	24.8
7	1.189	51.9%	51.9%	30.3%	783768534.7	27.8

Another way to check the efficiency in memory usage consists in measurements in coalescing global memory access on GPU [11] with the following two formulas [15]:

$$Read\ coalescent\ memory\ ratio = (l1\_global\_load\_hit + l1\_global\_load\_miss)/gld\_request \quad (1)$$
$$Write\ coalescent\ memory\ ratio = global\_store\_transaction/gst\_request \quad (2)$$

Where the nvprof events depicted in Equations 1 and 2 mean:

- $l1\_global\_load\_hit/l1\_global\_load\_miss$: Total number of access with/without success in data on the GPU L1 cache memory.
- $gld\_request/gst\_request$: Total requests to read/write in GPU global memory.
- $global\_store\_transaction$: Total number of store transactions in GPU global memory.

The metric value measured for read/write coalescing memory access should be close to 1.0 for 32-bit data ("float" type) [15]. Values greater than 1.0 indicate non-coalescing access. For version seven, the most efficient one, the metric values measured for reading and writing were 1.33 and 1.63, respectively, showing a non perfect coalescing memory access, which needs further investigations to identify the causes and to minimize the effects.

It is also possible to estimate a major bound for the algorithm developed, determining whether it is memory or computationally bounded. To measure this characteristic two extra versions of code were developed [15]. The first extra version performs only memory access operations without computation, and the second one only performs related logic and numerical operations instead of the original code, using the data already loaded into GPU registers. The result is shown in Figure 4, indicating an equivalent time spent on memory and computation only operations, characterizing a balanced algorithm with performance not limited by just memory access or numerical and logical procedures. So, future optimizations should focus not only in one aspect of algorithm.

Furthermore, Figure 4 also shows that the total advection time (column "full") is only slightly higher than the two measurements focusing only memory or computation, indicating a possible and beneficial overlapping between memory and logical/arithmetic operations. It also indicates that memory latency is not a significant performance boundary in this case.

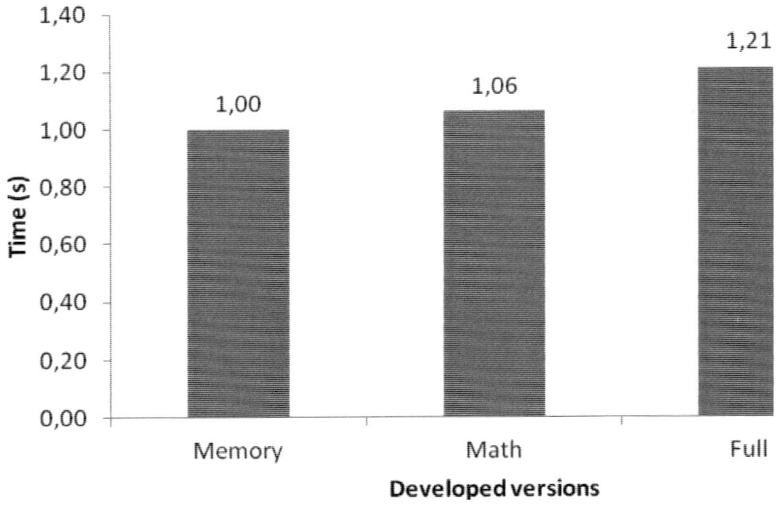

Figure 4. Memory / computation time spent

It was also possible to measure algorithm performance (for the best release: version seven) applied to data arranged in different coordinate axes of a three-dimensional computational grid, as shown in Table 4, considering the grid with the same dimensions mentioned configuration in Figure 3.

Table 4. Performance over coordinate axes

Coordinate Axis	Time (s)	Gigaflops
X	0.1026	66.6
Y	0.1037	65.4
Z	0.0998	70.0

According to Table 4 there was no significant performance difference between the data processed in different coordinate axes, despite the different memory arrangement of neighbor cells on each coordinate axis. The calculations over Z axis shows a slightly better performance comparing to others, which is expected, as the neighbor cells along Z axis are positioned closer to each other in the memory addresses when comparing to cell neighborhood in other axes, which suggests better memory access efficiency. However, it is important to highlight the differences in the number of tasks (and consequently in the number of streams), as well as their specific characteristics for each coordinate axis, which could interfere in performance.

The scalability of our method with respect to the problem domain size is shown in Figures 5 and 6. In Figure 5, the three-dimensional domain is configured to $168 * 168 * 38(XYZ)$ elements and the number of meteorological fields (N) ranges from 1 to 60 fields. In Figure 6, the number of meteorological fields and the number of points in vertical direction (Z axis) was set to 10 and 38, respectively, and the number of horizontal elements over X and Y axis ranges from 48 to 336 elements.

In Figure 5 it is clear to notice that the runtime does not increase linearly with the proportional increase in meteorological fields, since the estimated number of floating point

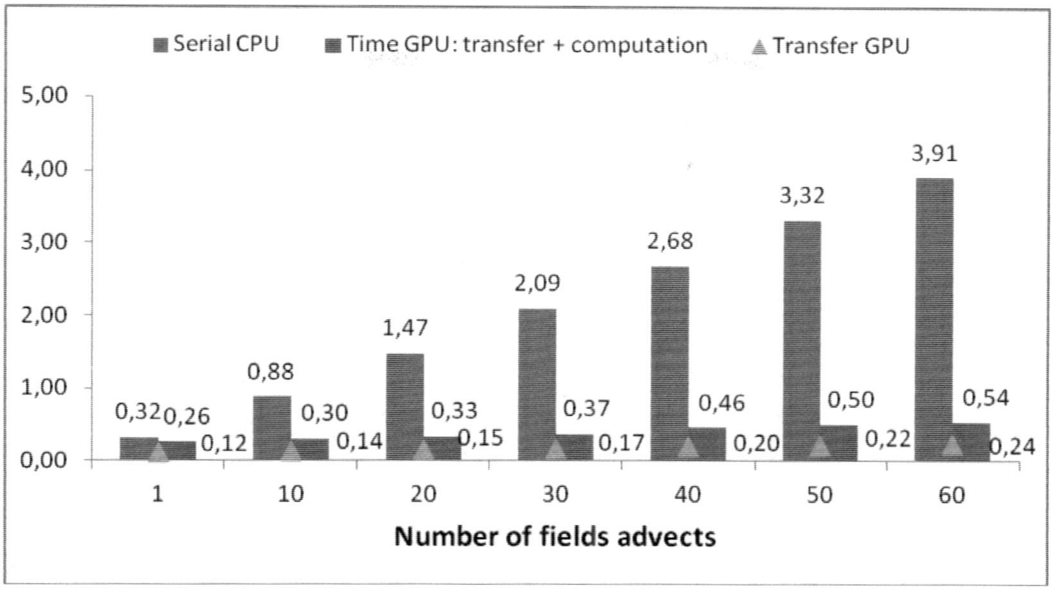

Figure 5. Scalability based on meteorological fields increasing

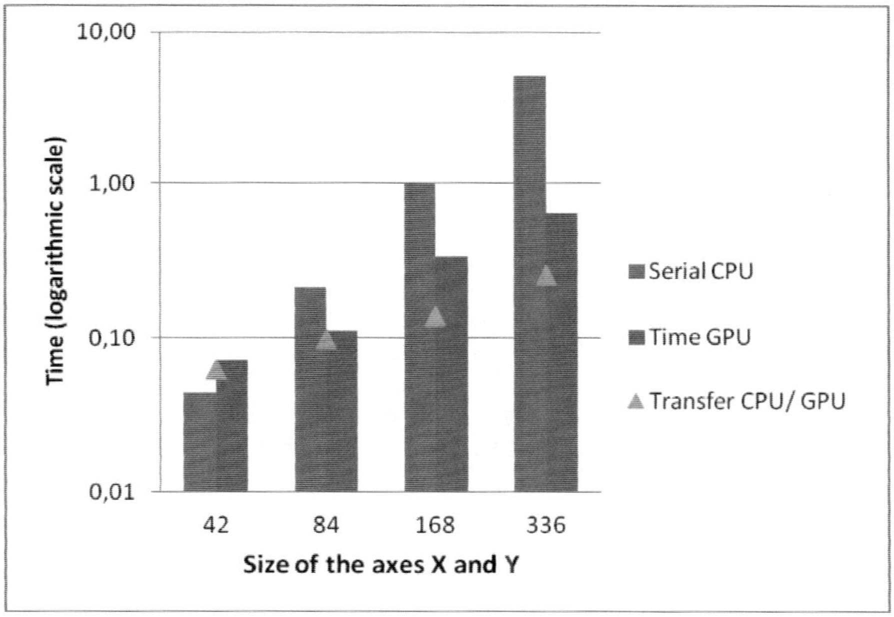

Figure 6. Scalability based on horizontal domain increasing

operations raises up to 60 times while runtime increases only 3.8 times. This could be confirmed observing the Gigaflops rate measured for the same case in Figure 7, which shows that the performance increases with the amount of data processed, reflecting a typical GPU behavior. It also reveals a non proportional transfer time increasing to a data volume growing.

Figure 6 shows a similar behavior as the execution time increase is not proportional to the continuous increase of 4 times for computational domain, implying in a computational effort

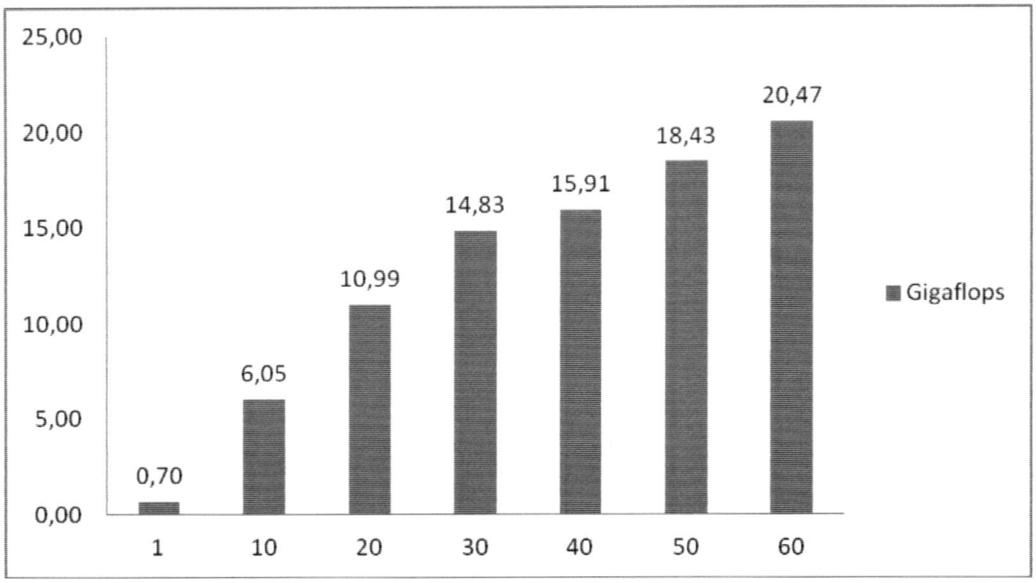

Figure 7. Scalability in Gigaflops based on meteorological fields increasing

increase of same order. Comparing the first and last domain size in same figure, it is possible to see a computational effort 64 times greater (42*24*38 and 336*336*38 respectively). However, the execution time grows to only 9.1 times the reference value. In the same figure one may observe the relative time spent on data transfers between host and device, which shows 85% of total time is due these transfers in lower domain but only 25% is spent for similar transfers for the greater domain. This shows an expected decreasing in data transfer importance related to increase of used data volume.

Comparing the scalability over a increasing number of meteorological fields (Figure 5) and over a increasing domain length (figure 6), the first case shows a better scalability, since the operations related to different meteorological fields are totally independent, generating more parallelism, while a domain variation could change the stream patterns, and the potential parallelism is dependent of the input data.

The robustness of the developed approach can be demonstrated using a more powerful accelerator and bigger problem domains. The results are shown in Table 5 for a computational domain size set as 60*38*336*336 ($NZXY$) in a machine configured with 12 cores *Intel Ivy Bridge* CPU, *NVidia Tesla K20X* accelerator and PGI Accelerator v.14.7 compiler, which shows GPU performance advantage comparing only up to 8 CPU cores. Using more CPU cores the OpenMP version reaches better results. It is worth to mention the significant data transfer time, which corresponds to 47% of total time. However, considering in the future a meteorological model fully ported to GPU acceleration, the cost of data transfer should be minimized since the input data could be already allocated on GPU memory originating from other part of code also running in same device. In this case the accelerated monotonic advection still has a performance 2 times better than a 12 cores CPU.

Another test was developed considering synthetic data instead of real input data like in previous tests, in order to verify the performance in a set of different input data configurations, which modify the number of concurrent stream tasks. The computational domain was set to 22*38*336*336 ($NXYZ$), and a synthetic data generation was applied and analyzed just for the data along X axis where the quantity of cells in left and right side of task streams

Journal of Physics: Conference Series **649** (2015) 012003 doi:10.1088/1742-6596/649/1/012003

(with different wind directions) was progressively increased, so, the stream size grows with a proportional decreasing in the total number of tasks to be processed. The results in Table 6 confirm an expected better performance with the maximum number of tasks, and consequent small size of streams, because it allows more parallelism. In real simulation, such situation occurs in cases where the wind direction changes frequently. Input cases where the wind has just a few variations, as it happens in high atmosphere, imply in long length stream and lower number of tasks, reducing the GPU performance.

Table 5. Performance with synthetic input data

Threads OpenMP or GPU	Time (s)	Speedup
1	12.03	1.0
2	6.33	1.9
4	3.36	3.6
6	2.44	4.9
8	1.84	6.5
10	1.44	8.4
12	1.30	9.3
GPU	1.49	8.1

Table 6. Results obtained with the same synthetic data size

Stream left	Stream right	Size of stream	Number of tasks	Time (s)
1	1	2	4004964	0.09
3	3	6	1347660	0.15
6	6	12	674316	0.15
18	18	36	241452	0.15
54	54	108	108782	0.15
162	162	324	60686	0.17

4.1. Inefficient versions

The last three versions presented in the beginning of this Section show no performance gains when compared to the final efficient version (release seven). The first one (release eight) consists of a new code where the task list is sorted according to both the stream type and size, trying to avoid divergent branches within a GPU Warp. However, this action did not generate significant improvements in performance. For a problem configured in the same way as described in Figure 4 (22*38*336*336 - $NZYX$) with real input data, the runtime was similar to that obtained without tasks ordering, considering the advection time plus the overhead necessary to sort the task list.

The second inefficient version works with hybrid processing, where both CPU and GPU work together and simultaneously to obtain the final result. The algorithm is designed allowing asynchronous GPU kernel launches overlapped with CPU processing. The GPU tasks are taken from the beginning of tasks list and the CPU tasks are taken from tail, in opposite direction. The designed algorithm was inspired in the OpenMP version (release two) and allows dynamic load balancing. However, the processing time shows just slightly improvements in performance (2% gain). The main reason to justify why the hybrid version does not show a significantly gain over the best computational time obtained rely on data transfers elapsed time. Since we are dividing the discrete domain between streams (or tasks) to be evaluated by CPUs and GPU, according the algorithm depicted in Listing 1, it is not possible to transfer to/from GPU just the portion related to some streams, because it represent a irregular domain over the 4D array. So, all domain data are transferred taking about 70% of OpenMP total time. The remaining time is insufficient to receive significant contributions by CPUs, since any stream processed simultaneously by 12 CPUs spend about 2.6 times than GPU (considering just Kernel execution time). Besides, it is necessary to post-process the data merging the results coming from different devices, which takes more time. The Listing 1 shows the current developed hybrid version. A new version where the domain is divided by atmospheric fields to be processed by CPUs and GPU are in progress. With this new version will be possible to transfer to/from GPU only the portion used on it.

Listing 1. Hybrid version

```
!$OMP PARALLEL &
!$OMP shared(ini, end, task_remain, chunk, i) &
!$OMP private(j, kernel_running)
do while (task_remain>0)

   ! Check if it is possible to launch a new GPU kernel
   kernel_running = .false.
   !$OMP CRITICAL
   if (acc_async_test_all()) then !kernel in execution?
      ! GPU is free
      call kernel_gpu_openacc(...)
      ini = min(ini+chunk, (TotalTasks - 1))
   else
      ! update total tasks remaining
      end = end - 1
      task_remain = end - ini + 1
      kernel_running = .true.
   endif
   !$OMP END CRITICAL

   ! Check if it is possible to launch a CPU thread
   if (task_remain>0 .and. kernel_running) then
      ! new CPU thread
      call process_advec_cpu(...)
   endif

enddo
!$OMP END PARALLEL
```

The motivation to create the last version (release ten) comes from an analysis of the source-code where it is possible to identify independent arithmetic operations by each grid cell. Therefore, a new version was developed joining two independent phases of the code: in the first phase a new numerical algorithm calculates everything that is possible to perform with independent parallel floating point operations by each grid cell, and the second phase uses the data produced by the first one as input to evaluate the advection following the depicted data-flow approach. The results show a second phase solving the advection phenomena in 77% of total time, but the algorithm for the first phase presents a performance higher than 23% of total time, so there is no performance gain. This code also requires significantly more GPU memory to store temporary data for the first part of code.

5. Discussions and Conclusions

The naive introduction of OpenAcc directives was not sufficient to achieve performance efficiency in GPU for the monotonic advection. It required several modifications such as algorithmic modifications and complementary OpenAcc directives applied to the original source-code, which improved the performance in 14.6 times since the first version in GPU. It was possible to drop the wall clock from about 3.33s (1.9 Gflops) to the final efficient version running the same problem with wall clock of about 0.74s (27.8 GFlops). Auxiliary tests also show balancing between the time spent on memory access and floating point operations in the final version, and a beneficial overlapping of them.

However, the amount of code changes and the programmer's effort were quite significant, which raises doubts about the expected high productivity of applying OpenACC directives [9] over original code versus the traditional CUDA language development; furthermore, there are some evidence related on this kind of problem in literature [10].

The performance gain with GPUs may be considered only modest compared to traditional optimizations advection methods [18, 21], where performance reaches hundreds of Gflops in GPUs. However, in this work the solution adopted to deal with flow dependence generates a coarse grain parallelism for GPU, due to grouping sets of data into streams, which needs to be evaluated serially. To exemplify the coarse grain parallelism, may be considered one case where the code developed uses approximately only 230,000 independent tasks and a few dozen dependent tasks to evaluate monotonic advection over a three-dimensional domain set to 38*336*336 cells (ZXY), which has more than 4 million cells. Thus, on average, each task is responsible for calculating the advection for 56 adjacent cells, while traditional numerical advection methods usually addresses concurrent tasks for each computational cell, generating much more parallelism.

The data access in GPU global memory for the developed code shows imperfect coalescing. Such effect may become the subject for future investigations trying to identify its causes and propose optimizations in order to minimize the effects.

It is also important to note that the gain related to CPU has the same order in different architectures used. Hybrid versions, which involve simultaneous CPU and GPU computation, did not improve performance significantly for the characteristics of the investigated problems.

Future developments to expand this work include, in addition to the aforementioned improvements in coalescing data access, code extensions allowing multiple GPUs in conjunction with distributed processing via MPI programming. Benchmarks considering productivity and performance using OpenACC and CUDA could also be addressed, as well as using other known accelerator as the *Intel Xeon Phi*.

References

[1] Boulos V, Huet V, Fristot V, Salvo L and Houzet D 2012 *Efficient implementation of data flow graphs on multi-gpu clusters.* Journal of Real-Time Image Processing 9 217-232
[2] Cumming C, Osuna T, Gysi T, Bianco M, Lapillonne X, Fuhrer O and Schulthess T C 2013 *A review of the challenges and results of refactoring the community climate code cosmo for hybrid cray hpc systems.* Proceedings of Cray User Group
[3] DocsNvidia 2014 *Profiler User's Guide CUDA Toolkit documentation*

[4] Fonseca R M, et al. 2012 *Dataflow paradigm approach applied to improve computational efficiency of a monotonic advection scheme.* Proceedings of Workshop PADTempo/XVII Congresso Brasileiro de Meteorologia. Gramado

[5] Freitas S R, Rodrigues L F, Longo K M and Panetta J 2011 *Impact of a monotonic advection scheme with low numerical diffusion on transport modeling of emissions from biomass burning. Journal of Advances in Modeling Earth Systems* 3

[6] Govett M, Middlecoff J and Henderson T 2010 *Running the NIM Next-Generation Weather Model on GPUs.* 10th IEEE/ACM International Conference on Cluster, Cloud and Grid Computing (CCGrid) 792-796

[7] Hartley T D R, Saule E and Catalyurek U V 2010 *Automatic dataflow application tuning for heterogeneous systems.* International Conference on High Performance Computing (HiPC)

[8] Haines W A 2013 *Acceleration of the Weather Research Forecasting (WRF) Model using OpenAcc and Case Study of the August 2012 Great Arctic Cyclone.* Master's thesis The Ohio State University

[9] Herdman J A, Gaudin W P, McIntosh-Smith S, Boulton M, Beckingsale D A, Mallinson A C and Jarvis S A 2012 *Accelerating Hydrocodes with OpenACC, OpeCL and CUDA.* SC Companion: High Performance Computing, Networking, Storage and Analysis (SCC) 465-471

[10] Hoshino T, Maruyama N, Matsuoka S and Takaki R 2013 *CUDA vs OpenACC: Performance Case Studies with Kernel Benchmarks and a Memory-Bound CFD Application.* 13th IEEE/ACM International Symposium on Cluster, Cloud and Grid Computing (CCGrid) 136-143

[11] Kirk D and Hwu W W 2010 *Programming Massively Parallel Processors: a hands-on approach.* Morgan Kaufmann 288

[12] Longo K M, Freitas S R, Pirre M, Marcal V, Rodrigues L F, Alonso M F and Mello R 2013 *The Chemistry CATT-BRAMS model (CCATT-BRAMS 4.5): a regional atmospheric model system for integrated air quality and weather forecasting and research.* Geoscientific Model Development 6 1389-1405

[13] Meng J, Morozov V A, Vishwanath V and Kumaran K 2012 *Dataflow-driven gpu performance projection for multi-kernel transformations.* SC '12 Proceedings of the International Conference on High Performance Computing, Networking, Storage and Analysis

[14] Michalakes J and Vachharajani M 2008 *GPU acceleration of numerical weather prediction.* IEEE International Symposium on Parallel and Distributed Processing, IPDPS 1-7

[15] Micikevicius P 2012 *GPU Performance Analysis and Optimization.* GTC2012

[16] OpenACC-Standard.org 2013 *The OpenACC Application Programming Interface.* OpenACC.org

[17] Shi S 2012 *GPU Implementation of Iterative Solvers in Numerical Weather Predicting Models.* Master's thesis The University of Edinburgh

[18] Souza A F, Veronese L, Lima L M, Badue C and Catabriga L 2012 *Evaluation of two parallel finite element implementations of the time-dependent advection diffusion problem: Gpu versus cluster considering time and energy consumption.* High Performance Computing for Computational Science - VECPAR 149-162

[19] Walcek C J *Minor flux adjustment near mixing ratio extremes for simplified yet highly accurate monotonic calculation of tracer advection.* Journal of Geophysical Research: Atmospheres, 105:93359348, 2000.

[20] Wang L, Shen C, Seetharaman G, Palaniappa K and Bhattacharyy S S 2012 *Multidimensional dataflow graph modeling and mapping for efficient gpu implementation.* IEEE Workshop on Signal Proc. Systems (SiPS) 300-305

[21] Wang Z, Xu X, Xiong N, Yang L T and Zhao W 2011 *GPU Acceleration for GRAPES Meteorological Model.* 13th IEEE International Conference on High Performance Computing and Communications (HPC) 365-372

[22] White J B and Dongarra J J 2011 *Overlapping computation and communication for advection on hybrid parallel computers.* IEEE International Parallel Distributed Processing Symposium (IPDPS) 59-67

XV Brazilian Symposium on High Performance Computational Systems (WSCAD 2014) IOP Publishing
Journal of Physics: Conference Series **649** (2015) 012004 doi:10.1088/1742-6596/649/1/012004

Optimizing Quantum Simulation for Heterogeneous Computing: a Hadamard Transformation Study

Anderson B. de Avila[1], **Murilo F. Schumalfuss**[1], **Renata H. S. Reiser**[1], **Mauricio L. Pilla**[1], **Adriano K. Maron**[2]

[1] Centre for Technological Development – Federal University of Pelotas
Pelotas/RS - Brazil

[2] Department of Computer Science
University of Pittsburgh Sennott Square – Pittsburgh/PA, USA

E-mail: {abdavila,mfschumalfuss,reiser,pilla}@inf.ufpel.edu.br, akk48@pitt.edu

Abstract. The D-GM execution environment improves distributed simulation of quantum algorithms in heterogeneous computing environments comprising both multi-core CPUs and GPUs. The main contribution of this work consists in the optimization of the environment VirD-GM, conceived in three steps: (*i*) the theoretical studies and implementation of the abstractions of the Mixed Partial Process defined in the qGM model, focusing on the reduction of the memory consumption regarding multidimensional QTs; (*ii*) the distributed/parallel implementation of such abstractions allowing its execution on clusters of GPUs; (*iii*) and optimizations that predict multiplications by zero-value of the quantum states/transformations, implying reduction in the number of computations. The results obtained in this work embrace the distribute/parallel simulation of Hadamard gates up to 21 qubits, showing scalability with the increase in the number of computing nodes.

1. Introduction

Although quantum computers are restricted to specific research centers and laboratories, the results from quantum information and quantum computation (QC) are a reality nowadays, and the first steps towards quantum hardware have been an important research subject [3, 4].

In many aspects, the foundations of QC are still seen as inaccessible or difficult as well as an instigating subject by congregating areas such as Physics, Mathematics, Electrical and Electronics Engineering. Additionally, QC includes relevant research in Computer Science and Computer Engineering by providing new concepts which can be improved with the study of quantum algorithms.

The investigation of properties such as superposition and entanglement leads to the development of new mathematical models and, consequently, quantum algorithms may be designed through the mathematical description of the system and by using simulation software. Moreover, its ideas may be applied to stimulate thinking about new mathematical tools and to solve open problems in conventional Computer Science.

The simulation of quantum algorithms in classic computers allows for development and testing that would otherwise be impossible without proper quantum hardware. However, there are many aspects of simulating quantum computers that are still open research topics, mainly due to the high temporal and spatial complexity of emulating quantum phenomena without proper optimizations, which restricts unoptimized simulation to a few quantum bits, or qubits.

The explosion of states and operations required to simulate quantum algorithms with even a small number of qubits seem to be a good match for manycore architectures. The current available architectures such as general-purpose computing on graphics processing units (GPUs) present a high number of processing elements well-suited for the regular operations found in quantum computing simulation, but the memory hierarchy in current designs is limited and is not usually transparent, i.e. the programmer must manage allocation of variables to different kinds of memory to achieve performance.

Further improvements in the complexity of simulated algorithms can be severely hampered by memory constraints, and thus adding distributed resources to the simulation becomes a necessity.

1.1. Research Problem

Composition and synchronization of quantum operators enable the exploration of parallel and distributed computations to model the quantum parallelism. However, the spatial and temporal complexity of simulating quantum computing increases exponentially with the number of qubits and operations. Due to such behavior, implementations based only on vector and matrix notations are not practical for even a small numbers of qubits.

In our project, we intend to integrate distributed and GPU-based simulation in a single environment, the Distributed Geometric Machine (D-GM).

In previous works, optimizations for an efficient representation of multidimensional Quantum Transformantions (QTs) were investigated in the D-GM environment [1, 15, 10, 11] . They contemplated the specification and implementation of D-GM environment and its main components, VPE-qGM and VirD-GM, enabling the modeling and distributed simulation of quantum algorithms, showing the evolution of quantum systems from a set of graphical interfaces.

1.2. Main Contribution

The general proposal to be achieved with the D-GM environment is **the development of new strategies for simulation of quantum algorithms, focusing on the reduction of the spatial and temporal complexity by considering partial processes in the definition of quantum transformations**.

In addition, based on the partiality concept, the granularity of distributed/parallel computation is investigated by considering classes of partial processes:

(i) the Classical Partial Processes (CPPs), related to partitions of read memory, which are represented by columns of its matrix representation;

(ii) the Quantum Partial Processes (QPPs), considering the write memory in the global memory state, represented by rows of the corresponding matrix representation; and

(iii) the Mixed Partial Processes (MPPs), obtained by integration of the two concepts above.

In this sense, the results contribute to integrate an efficient scheduling with more opportunity for performance gains, and new methods to balance communication and synchronization.

In particular, **this work aims to extend the possibilities of environmental simulation capabilities of the D-GM, introducing optimizations for representation and simulation of parallel/distributed quantum transformations** conceived in three steps:

(i) the theoretical studies and modeling of the abstractions of the MPPs, focusing on the reduction of the memory footprint for multidimensional quantum transformations;

(ii) the distributed/parallel implementation of such abstractions, allowing their execution on clusters of heterogeneous (non-uniform) computers with GPUs; and

(iii) optimizations that predict multiplications by zero-value amplitudes of the quantum state or quantum transformations which avoid unnecessary operations were developed, implying reduction in the number of computations.

The results achieved with the development of this work contribute for the consolidation of a more general objective – quantum simulations performing on hybrid computational system in the presence of hardware heterogeneity.

1.3. Paper Organization

This paper is structured as follows: Section 2 comprises the basics of Quantum Computing (QC). Section 3 describes the state-of-the-art in QC simulation. Afterwards, Section 4 presents an overview of the D-GM environment showing how computation is divided in the VirD-GM environment. Section 5 presents the modeling of modifications of VirD-GM environment and corresponding optimizations implemented in the current work. Section 6 details the parallel execution in the VirD-GM environment. Results are discussed in Section 7. Finally, conclusions and future work are drawn in Section 8.

2. Foundations of Quantum Computing

In QC, the *qubit* is the basic unit of information, defined by an unitary and bi-dimensional state vector that can be described in the Dirac's notation [12] by

$$|\psi\rangle = \alpha|0\rangle + \beta|1\rangle. \tag{1}$$

The coefficients α and β are complex numbers corresponding to the amplitudes of the respective states, respecting the normalization condition $|\alpha|^2 + |\beta|^2 = 1$ and guaranteeing the unity of the system state vector represented by $(\alpha, \beta)^t$. The amplitudes allow the representation of distinct states at the same time, configuring quantum superposition states that originate the quantum parallelism.

The state space of a quantum system with multiple *qubits* is obtained by the tensor product of the space states of its subsystems. Considering a quantum system with two *qubits*, $|\psi\rangle = \alpha|0\rangle + \beta|1\rangle$ and $|\varphi\rangle = \gamma|0\rangle + \delta|1\rangle$, the state space comprehends the tensor product $|\psi\rangle \otimes |\varphi\rangle$, described by

$$|\psi\rangle \otimes |\varphi\rangle = \alpha \cdot \gamma|00\rangle + \alpha \cdot \delta|01\rangle + \beta \cdot \gamma|10\rangle + \beta \cdot \delta|11\rangle. \tag{2}$$

The state transition in an N-dimensional quantum system is performed by quantum transformations (QTs) defined by square matrices of order N (i.e., 2^N components since N is the number of *qubits* in the system). As an example, the matrix notation for the unitary QTs *Identity*, *Hadamard*, *Phase Gate*, *T Gate* and *Pauly X* transformations [12, 8] are, respectively, defined by

$$Id = \frac{1}{\sqrt{2}}\begin{pmatrix} 1 & 0 \\ 0 & 1 \end{pmatrix}, H = \frac{1}{\sqrt{2}}\begin{pmatrix} 1 & 1 \\ 1 & -1 \end{pmatrix}, S = \begin{pmatrix} 1 & 0 \\ 0 & j \end{pmatrix}, T = \begin{pmatrix} 1 & 0 \\ 0 & \exp^{\frac{j\pi}{8}} \end{pmatrix} \text{ and } X = \begin{pmatrix} 0 & 1 \\ 1 & 0 \end{pmatrix}.$$

The application of such transformations to a quantum state $|\psi\rangle$ generates the new corresponding global states $H|\psi\rangle = \frac{1}{\sqrt{2}}(\alpha + \beta)|0\rangle + \frac{1}{\sqrt{2}}(\alpha - \beta)|1\rangle$ and $X|\psi\rangle = \beta|0\rangle + \alpha|1\rangle$.

Additionally, QTs simultaneously applied to different *qubits* imply in the tensor product of the corresponding matrices, as described for the Hadamard transformation in Eq. (3):

$$H^{\otimes 2} = \frac{1}{\sqrt{2}}\begin{pmatrix} 1 & 1 \\ 1 & -1 \end{pmatrix} \otimes \frac{1}{\sqrt{2}}\begin{pmatrix} 1 & 1 \\ 1 & -1 \end{pmatrix} = \frac{1}{2}\begin{pmatrix} 1 & 1 & 1 & 1 \\ 1 & -1 & 1 & -1 \\ 1 & 1 & -1 & -1 \\ 1 & -1 & -1 & 1 \end{pmatrix} = \frac{1}{\sqrt{2}}\begin{pmatrix} H & H \\ H & -H \end{pmatrix}. \tag{3}$$

Besides the unitary transformations, controlled transformations are also preserved by the tensor product, modifying the state of one or more qubits considering the current state of other qubits in a multidimensional quantum state [14].

The CNOT quantum transformation receives the tensor product of two qubits $|\psi\rangle$ and $|\varphi\rangle$ as input and applies the NOT ($Pauly\ X$) transformation to one of them (target qubit), considering the current state of the other (control):

$$CNOT \;=\; \begin{pmatrix} Id & 0 \\ 0 & X \end{pmatrix}. \tag{4}$$

The CNOT controlled transformation applied to a generic two-dimensional superposition quantum state $|\psi\rangle \otimes |\varphi\rangle$ results in the following quantum state:

$$CNOT\,(|\psi\rangle \otimes |\varphi\rangle) = \alpha \cdot \gamma|00\rangle + \alpha \cdot \delta|01\rangle + \beta \cdot \delta|10\rangle + \beta \cdot \gamma|11\rangle. \tag{5}$$

Controlled transformations can be generalized in a similar way, e.g. the Tofolli and Fredkin transformations [12, 7].

3. Related Work

There is a number of parallel simulators trying to tackle the complex issues with efficient quantum computing simulator. Most approaches use clusters or GPUs to achieve reasonable performance and system complexity with different results. We discuss the most relevant works in the next subsections.

3.1. Quantum Simulation in Supercomputers

The *Massive Parallel Quantum Computer Simulator* (MPQCS) [13] uses the MPI library for communication, facilitating its execution on many parallel machines with or without shared memory. Quantum algorithms are described from a universal set of QTs such as $\{H, S, T, CNOT\}$. Combinations of these transformations may describe any quantum algorithm. This design choice promotes optimizations in the simulator, as the required calculations are predictable. However, designing and implementing quantum algorithms with higher complexity is bound to suffer from the lack of higher abstractions.

The most recent publication in a similar environment dates from 2010 [6], where the simulation of the Shor algorithm with 42 qubits was executed on the JUGENE supercomputer and the number 15707 was factored as 113×139. This simulation required 262,144 processors, without specifing memory usage and execution time.

3.2. GPU Quantum Simulation

Gutierrez et al. [5] used the CUDA framework to explore the parallel nature of quantum algorithms. In their approach, calculations required for the evolution of the quantum system are executed by thousands of threads in a GPU. A set of QTs of one and two qubits are more expressive than those found in Raedt et al. [13], however, the bigger the set of QTs, the more complex is the simulation algorithm due to the variety of patterns that are possible in the definition of a multi qubit transformation.

The main limitation with GPU approaches is related to the memory capacity, which are usually restricted to a few gigabytes. Gutierrez et al. [5] were able to simulate systems with 26 qubits at most with the 768 MB available in the GeForce 8800 GTX. When compared to extremely optimized simulations in general-purpose CPUs, speedups of 95 times are possible.

3.3. Hybrid Quantum Simulation
Advances in the programming environments and capabilities of both GPUs and CPUs resulted in a new tendency towards exploration of heterogeneous resources to further improve efficiency. Therefore, many applications are being adapted to take advantage of these new architectures. As far as the authors know, ours is the first integrated and hybrid environment for simulation of quantum computing.

Although the advantages of such hybrid environments are clear, there are many challenges related to communication, programming, and synchronization that must be overcome to make it feasible. There are two main justifications for facing these challenges:

(i) the computational power found in clusters with GPUs is much greater than those with only CPUs;

(ii) the cost/benefit (both in terms of performance and energy consumption) for applications well-suited for GPUs is much better than for CPUs.

In previous works, we presented how our approach can be used for computing of the Hadamard Transformation [9, 2] in a GPU environment. In the current work, we extend our previous work to improve efficiency and scalability to the quantum applications performed up to now on heterogeneous platforms and further on hybrid platforms.

4. GPU-Aware Distributed Quantum Simulation
The *Distributed Geometric Machine* (D-GM) project proposes a complete framework for simulation of quantum algorithms. Different graphical interfaces provide tools for modeling the application and simulating it, either sequentially or in parallel, using GPUs or multicore CPUs.

Figure 1 illustrates how the *D-GM Simulation Framework* is organized. Each different level has its own purpose towards the unification of the modeling/simulation process. In the *Quantum Circuit Level*, one can describe the application following the quantum circuit model and then automatically export it to a corresponding representation in the qGM (Quantum Geometric Machine) model [11]. The qGM level contains the *VPE-qGM* (Visual Programming Environment for the qGM Model) environment that allows the user to describe and simulate the computations under the qGM model. In the D-GM level, the distributed simulation manager, VirD-GM (Virtual Distributed Geometric Machine) – seamlessly handles tasks such as communication, scheduling, and synchronization when a distributed simulation is required. Finally, the hardware level enlists all the devices that can be explored by this framework, from regular desktops for sequential simulation to clusters with multiple GPUs.

The VirD-GM environment [9, 2] provides transparent management of the distributed execution for the *VPE-qGM* environment.

Distributed execution of applications from VirD-GM requires the following inputs: *(a)* a process description file, *(b)* a memory description file, and *(c)* the addresses of the execution nodes.

The main modules found in the VirD-GM environment are:

(i) **VirD-Loader**: responsible for the interpretation of the input files;

(ii) **VirD-Launcher**: executes scheduling and synchronization of tasks; and

(iii) **VirD-Executor**: controls communication and data transfer among execution nodes (VirD-Clients).

Currently, there are three distribution options provided by the D-GM environment that allow for the definition of different computing layouts for the same quantum transformation which maintain a coherent quantum interpretation for the simulation of total/partial transformations,

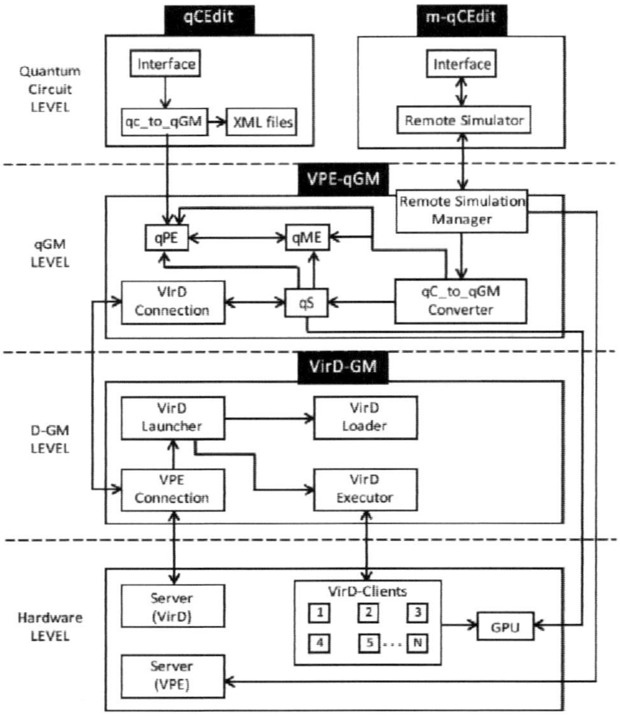

Figure 1. The D-GM Simulation Framework

as well as total and partial states. Each computing layout defines a method to divide and execute the operations that are required for a complete quantum computing simulation. They are described in the next subsections.

4.1. Distribution of Quantum Partial Processes

A *Quantum Partial Process* (QPP) [1] abstracts information about the rows of a matrix for a quantum transformation. Hence, only a few amplitudes from the current state vector are updated – those that were calculated from the rows that compose each QPP. The $\perp$ value in the rows denotes the omitted information. In this approach, the calculation of each QPP will be executed in an independent way for each set of amplitudes from the current state associated to the quantum transformation, without requiring shared data. The number of amplitudes updated in the current state vector after the execution of each QPP characterizes the granularity of the computation.

Figure 2 shows the division of a Quantum Process (QP) defining a Hadamard transformation of second order, $H^{\otimes 2}$, in QPPs. The first column depicts the original transformation in terms of the geometric model, as well as the matrix representing it. The second column shows the same calculations divided in two QPPs, where each QPP executes the operation for 2 amplitudes of the state vector. Granularity is represented by the tuple $(2, 4)$, i.e. 2 rows are being updated and 4 amplitudes from the global state are being read. The last column shows a set of QPPs with granularity $(1, 4)$, which are the most elementary QPPs possible for this example.

QPPs may be associated to nodes in a distributed system but each of them must have an updated copy of the current state vector for the entire system. This implies in the need of an expanded amount of memory in each node as well as heavy communication before each set of

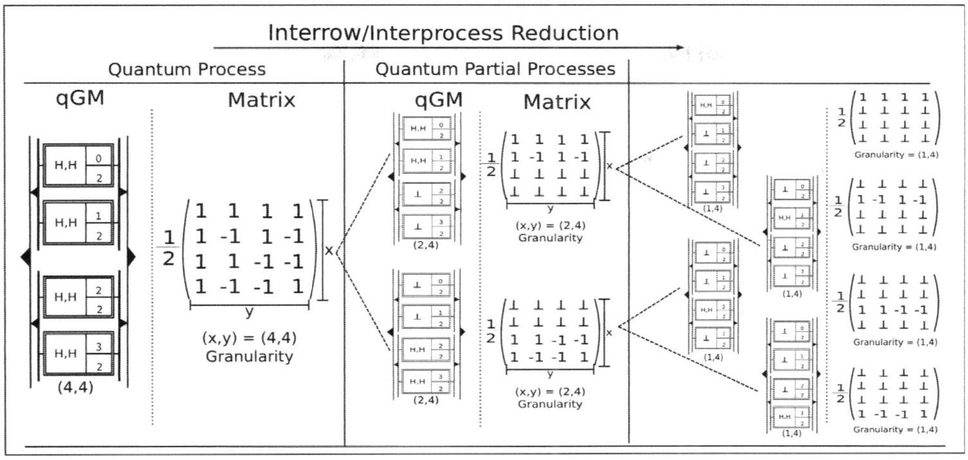

Figure 2. Distribution of the $H^{\otimes 2}$ operator among *QPPs.*

transformations.

4.2. Distribution of Classic Partial Processes
The distribution of *Classic Partial Processes* (CPPs) [9] allows for the abstraction of information about the columns of the matrix notation in a QT. Partial results are generated from the CPPs for all amplitudes of the resulting state vector.

Figure 3 shows some of the possible distribution of CPPs for the Hadamard operator $H^{\otimes 2}$. The $\perp$ value is generated in the columns that will be omitted according to the proposed abstraction.

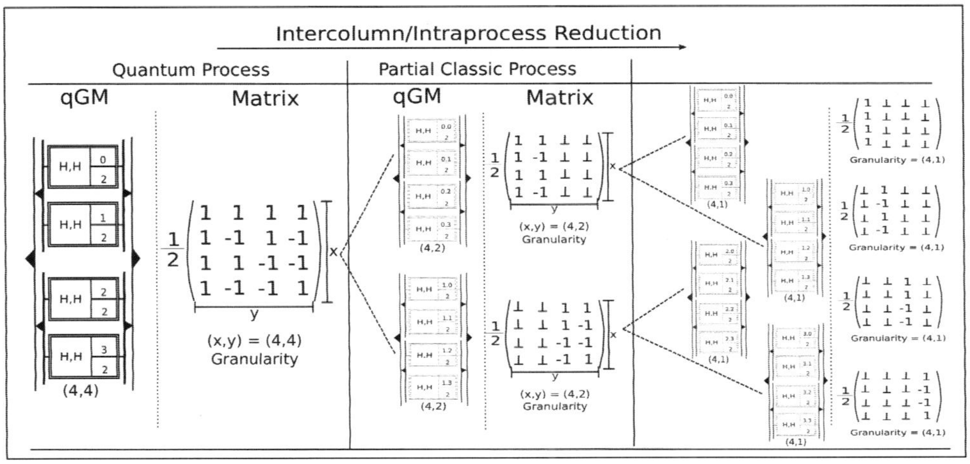

Figure 3. Distribution of the computations for the $H^{\otimes 2}$ transformation in CPPs.

In an analogous way to the described in Figure 2, the process depicted in Figure 3 presents its coarser granularity defined by $(4, 1)$, and each synchronization will generate a partial amplitude for each state in the computational base. The computation of each CPP is independent from the others and may be executed in parallel if there are enough resources to keep more than one partial amplitude.

The CPPs can also be associated to nodes in a distributed system, where each node receives part of the input state. With this information, it is possible to start calculating CPPs associated to a QT. However, each CPP generates an entire output state vector, with partial values for the results, which increases memory consumption in the nodes.

4.3. Mixed Distribution

Both QPP Distribution and CPP Distribution have their advantages and disadvantages. The former requires the storage of only a few amplitudes, at the cost of having the entire current state space for the computation of its subset of amplitudes. The latter does not demand all amplitudes of the current state vector, although an equivalent memory space is necessary to store the partial amplitudes calculated by each CPP.

Our proposed combination of the QPP and CPP distributions in a mixed layout addresses both challenges: managing the communication overhead and reducing the memory usage. This enhanced layout is named *Mixed Partial Process* (MPP).

The notion of partial process corresponding to a MPP deals with the partition of read/write memory values considering a pair of parameters (R, W). Thus, the representation of a QT demands $R \times W$ MPPs, with R and W indicating the partition number of the read and write memory, respectively.

By the definition of a MPP, it is possible to integrate both interpretations:

(i) the QPP behaviour, by using a $(1, W)$ MPP configuration where only the write memory partition is considered; and

(ii) the CPP behaviour, by using a $(R, 1)$ MPP configuration where only the read memory partition is considered.

With the modeling and implementation of MPPs in VirD-GM, programmers receive more control of the memory usage and the granularity of computation. The integration of these features with the VPE-qGM interfaces provides for a more interactive experience with the simulations in the D-GM environment. Although the environment is still much higher-level than the interfaces provided by MPI and CUDA, the programmer retains the capability of fine-tunning the amount of calculations and memory usage according to resources available in the processing nodes.

Figure 4 shows an example of computation layouts in the MPP approach for a Hadamard $H^{\otimes 2}$ operator. In this context, the application of the mixed reduction in the associated matrix will reduce granularity from $(4, 4)$ to $(2, 2)$ in the distribution of the corresponding computation.

Figure 5 depicts the distribution of the computation among nodes:

(i) **Node 1** executes a calculation over half the amplitudes from the state vector, thus generating the partial amplitudes for the corresponding positions in the state vector; meanwhile,

(ii) **Node 2** executes the complementary calculation.

In the end, the partial amplitudes from each state are merged in the server node to obtain the final state for the quantum system using operators such as composition, sum, and others.

5. VirD-GM and implementations

The VirD-GM environment is being developed in order to produce the transparent management of distributed execution in the *VPE-qGM*. This section describes the control-operation of the VirD-GM environment to support the mixed modeling computing and implementations of MPPs.

Distributed simulation of a quantum algorithm from MPPs consists of sending every MPP to a *VirD-Client*, passing the parameters of the MPP and the necessary portion of the current

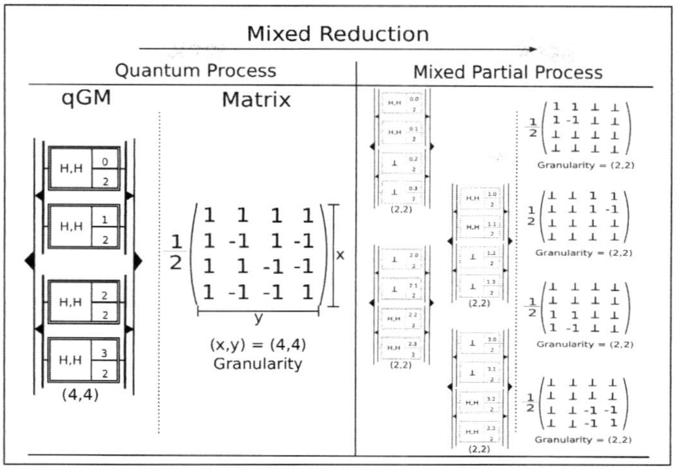

Figure 4. Mixed distribution of the $H^{\otimes 2}$ operator.

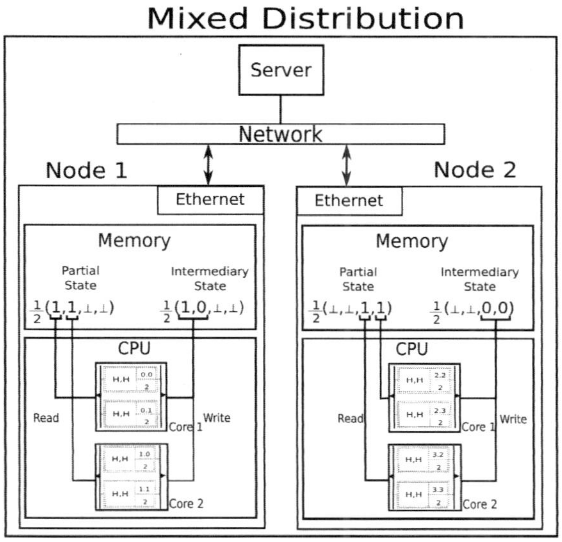

Figure 5. Example of the mixed distribution in a cluster of workstations.

memory. The *VirD-Client* performs the computation and generates a list with the calculated partial values, which is then sent back to the *VirdServer*. Afterwards, the *VirdServer* updates its memory with the data in this list.

This process is executed for all temporary files generated in the simulation, requiring a file for each step of the simulation. After all steps have been performed, the results can be exported to the VPE-qGM, which can load the simulation results and present them at the graphical interfaces of its editors.

5.1. Implementation

This implementation aims to contribute to reducing spatial and time complexity in multi qubits simulations.

The strategy adopted to reduce the spatial complexity is based on that an MPP may be defined from basic matrices of lowest order, which can be dynamically combined to generate the elements corresponding to the resulting matrix of the tensor product. The rows and columns of a matrix notation in a QT modeled by MPPs define which state vector amplitudes are necessary to perform the computation and also which partial state will be generated from its execution.

Complementary MPPs (interpreting disjoint sets of rows and columns of the same matrix associated with a QT) can be grouped in the descriptor file processes in order to compute an entire QT.

To reduce time complexity, all computation involving input positions with zero or QT elements with zero are not executed because they do not change the final result. In order to know beforehand if a position is associated to a valid iteration (generate a non-zero value), logic operations are carried out using two masks, $ZERO$ and ONE, so there is no need to iterate until the point at which it becomes invalid to discover this.

The main changes that made possible the extension of VirD-GM environment to support these optimizations are briefly described next:

(i) VirD-Loader module: modified to load the simulation memory in a format compatible for communication with the GPU, since Java provides no primitive complex type.

(ii) Process descriptor file: modified for recognition and interpretation of MPPs.

(iii) Data structures: transformed *Complex* structures into *Float* structures so that the kernel can interpret two consecutive floating-point numbers as a single complex number.

(iv) Vird-Exec module: modified to send to a VirD-Client only the portion of memory the MPP requires for simulation.

(v) QGM-Analyzer library: responsible for most of the execution, modified to enable calling the CUDA kernel when requested by management to run on GPUs. The framework JCuda [16] is used for kernel invocation and communication with the GPUs.

(vi) CUDA kernel: modeling and implementation of a new CUDA Kernel, providing support to the new optimizations.

6. MPPs Parallel Execution

This section introduces the efforts which enable the distributed/parallel execution of MPPs using the VirD-GM environment.

6.1. Data structure

In order to increase the granularity of distributed computations, the MPPs are defined from basic matrices of smaller order, which are combined in an iterative manner to dynamically generate the elements corresponding to the resulting matrix of the tensor product. When executing an MPP, the runtime uses the following parameters:

(i) Matrices List: list of basic matrices generated by host-code;

(ii) Last Matrix: contains the data of the last basic matrix, and is stored separately from the others because it is constantly accessed.

(iii) Zero: contains the values of each row related to column Zero of each quantum transformation operators used to generate the mask Zero;

(iv) One: contains the values of each line related to column One of each quantum transformation operators used to generate the mask One;

(v) Begin: stores in which position of the matrices list is the first element of each matrix.

(vi) Dimension: stores the dimension of each basic matrix.

(viii) Exp: stores the dimension log of each basic matrix.

All values corresponding to a QT matrix are stored in a contiguous data vector and the information about the beginning of the matrices indicate which data belong to which matrix. Moreover, the way these are combined is detailed in the CUDA kernel presented in subsection 6.3.

6.2. Data Allocation on GPUs

The arrangement of data over the different GPU memory spaces is particularly important to ensure good performance. The MPP parameters do not change during the simulation and are common to all threads. Thus, the allocation of that data in the constant memory space becomes an appropriate option. Data copies are made using the JCuda framework.

In addition to the MPP data, three vectors are found in the host memory space: (1) one modeling the current state of the quantum system; (2) one stating for each position which is the next with a non-zero value; and (3) other modeling the next state, which will be calculated. Before the implementation of the CUDA kernel may start execution, both vectors are copied to the GPU global memory.

6.3. CUDA Kernel

The CUDA kernel is implemented in an iterative fashion, since most of the GPUs do not support recursion (only GPUs with Compute Capability 3.5 or higher support resursion). As the kernel described here presents a behavior similar to the tensor product it is able to act on an arbitrary number of basic matrices. Each CUDA thread maintains internal information and execution controls to set access limits in each array.

The CUDA kernel computation can be divided into eight steps, as described below.

Step 1: **Constant information** initialization, which are common to all the CUDA threads of the application. *BITS* equivalent to 2 raised to the power of number of MPP qubits, is used in calculation operations of the element position on matrix; *AMP* informs how many amplitudes each CUDA thread calculates; *SHIFT_READ* and *SHIFT_WRITE* define the MPP reading and writing position basis, respectively; *TAM_BLOCK* sets the block size; *TOTAL_ELEMENTS*, *LAST_MATRIX_ELEMENTS*, *STACK_SIZE* and *STACK_LINES* are the constant memory vectors parameters.

These constants are defined at run time by changing the kernel base file, where these values are not defined yet because this information varies from execution to execution according to the MPP to be calculated, after that it is possible to compile the resulting kernel.

Step 2: **Initialization** of each thread's amplitude vector and local variables, depicted in Figure 6. The only value that differs among them is the *rowId*, which defines the thread row set.

```
for (l = 0; l < AMP; l++)
    newAmplitudes[l] = make_cuFloatComplex(0,0);

cuFloatComplex value = make_cuFloatComplex(1,0);
rowId = (TAM_BLOCK * blockIdx.x + threadIdx.x) * dimensionC[STACK_SIZE] + SHIFT_WRITE;
d = dim = prev_pos = pos = zero = one = 0;
```

Figure 6. Step 2: Initialization code executed by all threads.

Step 3: **Creation of masks** used to see if a position has a valid iteration in the thread row set, as shown in Figure 7.

Journal of Physics: Conference Series **649** (2015) 012004 doi:10.1088/1742-6596/649/1/012004

```
for (ind = 0 ; ind < (STACK_SIZE); ind++){
    mask = dimensionC[ind] -1;
    d += expC[ind];
    l = (rowId >> (BITS - d)) & mask;
    zero = (zero << expC[ind]) | (rowZeroC[dim + 1]);
    one = (one << expC[ind]) | (rowOneC[dim + 1]);
    dim += dimensionC[ind];
}
zero = (zero << LAST_EXP) | rowZeroC[dim];
one = (one << LAST_EXP) | rowOneC[dim];
```

Figure 7. Step 3: Code for masks.

Step 4: **Advance in the input vector** performing the calculations required for the positions that are different from zero and have a valid iteration (Figure 8).

```
p = positions[p];
while (p < SIZE){
    temp = readMemory[p];
    pos = p + SHIFT_READ;

    if (!((zero & ~pos) | (one & pos))){
    ...
```

Figure 8. Step 4: Calculation of valid iterations.

Step 5: **Backtrack the necessary iterations.** Based on the previous and current position, the variable used on the iterations has its value regressed to the point in common between the two positions (Figure 9).

```
back = (pos & (~prev_pos)) >> LAST_EXP;
while (back != 0){
    ind--;
    mask = dimensionC[ind] - 1;
    l = (rowId >> (BITS - d)) & mask;
    c = (prev_pos >> (BITS - d)) & mask;
    value = cuCdivf(value, matricesC[initMatrixC[ind] + l*dimensionC[ind] + c]);
    d -= expC[ind];
    back = back >> expC[ind];
}
```

Figure 9. Step 5: Backtracking a variable to a common point.

Step 6: **Advance in matrix** iteratively, calculating the multiplication between the elements indexed according to the *rowId* and current position values (Figure 10).

```
for ( ; ind < (STACK_SIZE); ind++){
    mask = dimensionC[ind] -1;
    d += expC[ind];
    l = (rowId >> (BITS - d)) & mask;
    c = (pos >> (BITS - d)) & mask;
    value = cuCmulf(value, matricesC[initMatrixC[ind] + l*dimensionC[ind] + c]);
}
```

Figure 10. Step 6: Calculation of the multiplication of elements indexed by *rowId*.

Step 7: **Perform the new amplitudes' partial update** using the value calculated on the previous step and going through the last matrix (Figure 11).

```
    c = pos & (AMP - 1);
    temp = cuCmulf(value, temp);
    for (l = 0; l < AMP; l++)
        newAmplitudes[l] = cuCaddf(newAmplitudes[l],
                                   cuCmulf(temp, lastMatrixC[l*AMP + c]));
    prev_pos = pos;
    }
  p = positions[p+1];
}
```

Figure 11. Step 7: Partial update of amplitudes.

Step 8: **Copy of vector data** with the new amplitudes calculated to the write state vector in GPU global memory (Figure 12

```
rowId -= SHIFT_WRITE;
for (l=0; l<AMP; l++)
    writeMemory[rowId+l] = newAmplitudes[l];
```

Figure 12. Step 8: Updating the global memory with the new amplitudes.

All these steps define the CUDA kernel and what each CUDA thread will execute recalculating n amplitudes of the new state vector according to the QTs applied, where n is equivalent to the size of the last matrix.

7. Evaluation

For validation and performance analysis of the simulation of quantum algorithms from MPPs in the VirD-GM, case studies with QTs Hadamard up to 21 qubits ($H^{\otimes 18}$, $H^{\otimes 19}$, $H^{\otimes 20}$ and $H^{\otimes 21}$) were considered.

The operator Hadamard was chosen because it corresponds to a QT (associated with dense matrix) that has the highest execution cost in the simulation environment.

The methodology adopted for the simulation distributed via GPU considers running 10 simulations of each instance of the operator Hadamard, considering each of the settings within the cluster of GPUs and three types of input state memory, the number of MPPs never exceeds the number of GPUs.

The tests were performed in four desktops with Intel Core i7 processors and 8 GB of RAM memory. Two desktops had NVIDIA GT640 GPUs, while the remaining two had NVIDIA GTX560 GPUs. An additional machine, connected to the cluster through a Fast Ethernet was requited to run the VirD-GM server. As software components, we have: JCUDA 0.5.5, NVIDIA CUDA Toolkit 5.5, and Ubuntu 12.04 64 bits. For simulations with 1 and 2 clients, the two desktops used were those with NVIDIA GT640.

The number of MPPs in each simulation is equal to the number of execution clients. Based on the pattern described in Section 4.3, related configurations were used as in the following:

(i) 1 VirD-Client: $(1-1)$;

(ii) 2 VirD-Client: $(1-2)$, $(2-1)$;

(iii) 4 VirD-Client: $(1-4)$, $(4-1)$, $(2-2)$.

Moreover, the memories used are reported in the following:

(i) Memory 1: all positions with non-zero values (superposition state);

(ii) Memory 2: non-zero and zero values alternated (intermediate state);

(iii) Memory 3: first position with a non-zero value and the rest with zeros values(classical state);

7.1. Results

Simulation times obtained can be seen in Tables 1, 2 and 3. In the simulations with memory 1 and 2, the standard deviation maximum of 0.79% was measured in the $H^{\otimes 18}$ with configuration $4-1$ using memory 2. In the simulations with memory 3, the standard deviation maximum of 4.9% was measured in the $H^{\otimes 20}$ with configuration $1-4$.

Table 1. Simulation average times obtained for memory 1, measured in seconds.

QT \ Conf.	1 *VirD-Client*	2 *VirD-Clients*		4 *VirD-Clients*		
	$1-1$	$1-2$	$2-1$	$2-2$	$1-4$	$4-1$
H^{18}	8.83	5.11	5.15	3.28	3.41	3.27
H^{19}	31.03	16.43	16.70	9.11	9.32	9.25
H^{20}	119.74	63.89	63.34	33.30	33.70	33.34
H^{21}	470.80	248.96	245.47	126.62	125.78	127.02

Table 2. Simulation average times obtained for memory 2, measured in seconds.

QT \ Conf.	1 *VirD-Client*	2 *VirD-Clients*		4 *VirD-Clients*		
	$1-1$	$1-2$	$2-1$	$2-2$	$1-4$	$4-1$
H^{18}	6.19	3.77	3.79	2.56	2.64	2.56
H^{19}	20.95	11.27	11.43	6.43	6.60	6.52
H^{20}	78.75	41.21	41.33	21.74	22.22	21.78
H^{21}	308.19	158.79	158.24	81.16	81.70	81.52

Table 3. Simulation average times obtained for memory 3, measured in seconds.

QT \ Conf.	1 *VirD-Client*	2 *VirD-Clients*		4 *VirD-Clients*		
	$1-1$	$1-2$	$2-1$	$2-2$	$1-4$	$4-1$
H^{18}	1.24	1.25	1.25	1.30	1.30	1.28
H^{19}	1.48	1.50	1.49	1.68	1.56	1.53
H^{20}	1.93	1.96	1.96	2.23	2.15	2.06
H^{21}	2.86	2.95	2.92	3.31	3.30	3.18

Figures 13, 14 and 15 present relative speedups obtained for the different types of configurations with the baseline being the $1-1$ configuration, which uses 1 VirD-Client. The relative speedup is the ratio between the baseline execution time and the parallel one:

$$Speedup_{parallel} = \frac{t_{baseline}}{t_{parallel}} \tag{6}$$

It can be observed that there is performance gain when increasing the number of clients for memories 1 and 2, regardless of the configuration used. For these cases, speedups tend to the ideal with the increased number of QTs qubits. This is because the time required

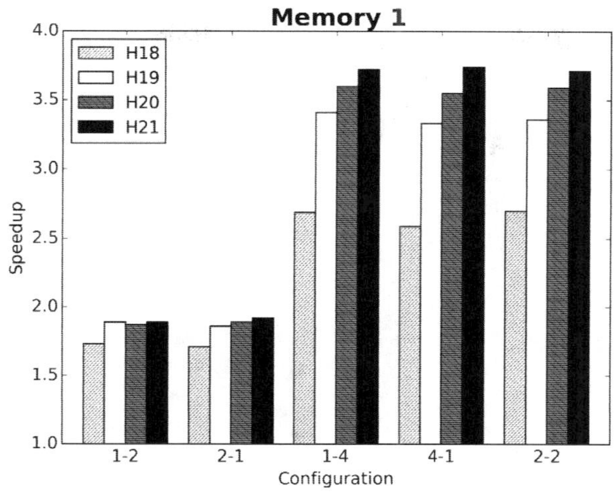

Figure 13. Speedup for Memory 1

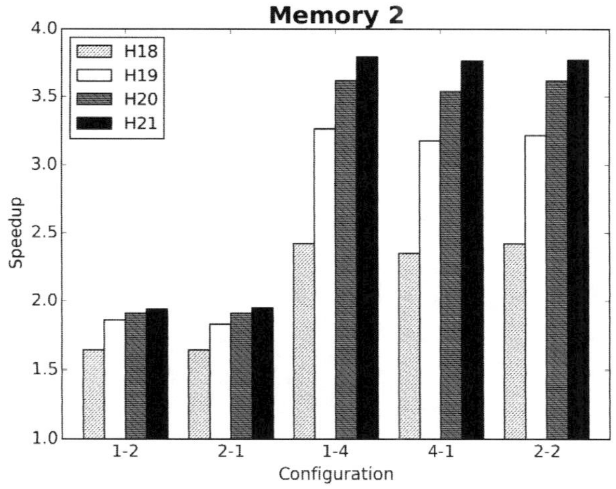

Figure 14. Speedup for Memory 2

for communication between the server and the client becomes less significant compared to the running time on the client.

Tests with the memory 3 (Figure 15) showed no performance gain because the time spent running is small, since there is only one valid position for computation, then the time spent on communication becomes the simulation bottleneck. Notice that this graph presents a different range of values in the vertical axis to better ilustrate the differences between bars.

In previous works, the main improvements in the D-GM environment were already introduced, using QPPs in order to control the granularity of computations in both approaches:

(i) distributed computations by using multicore CPU architecture in the VirD-GM [1], obtaining distributed simulation of systems comprised by Hadamard transformations from 14 to 17 qubits); and

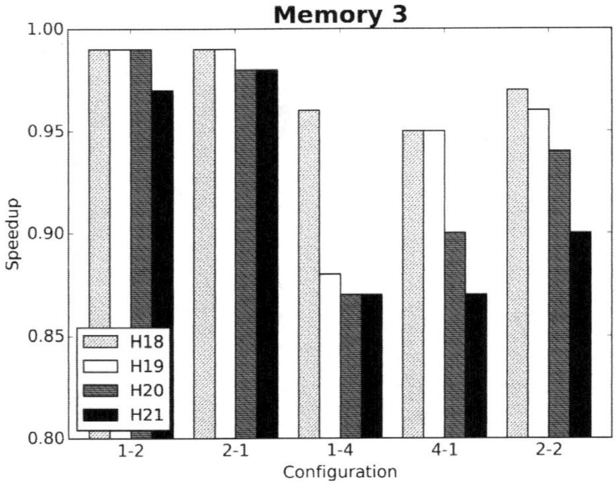

Figure 15. Speedup for Memory 3

(ii) massive parallel computations, by using multicore parallel/streaming processors provided by GPU arquitecture, in the extension of the qGM-Analyzer library which support executions in the VPE-qGM [9] (up to 20 qubits for Hadamard transformations performed over single GPU).

Based on these previous results, it is clear that there is a performance improvement obtained by QTs represented by dense matrices (as the Hadamard Transformations) supported by the present GPU approach in the VirD-GM (up to 21 qubits for Hadamard transformations performed over a GPU cluster) in comparison with the distributed simulation supported by multi-core CPU architecture in the VirD-GM.

Other QTs (as controls and measurement operators) are still being incorporated to multi-GPU units approach therefore unavailable in this step for comparative analysis.

8. Conclusion and Final Remarks

The extension of the VirD-GM environment described in this paper provides high performance and scalable simulation of quantum algorithms. We tackled the issues of high spatial complexity and execution time by the use of MPPs in a computing platform comprised of computers with manycore GPUs, integrated through a network.

The concept of a MPP and the read/write memory partition allow the granularity control of an application. Therefore, the GPU memory does not seem an important limitation in the increment of quantum application dimensions, increasing scalability.

As a case study, we simulated the Hadamard Transformation, a challenging quantum algorithm that stresses both spatial and temporal complexities.

The computations optimization for different memory types and the development of the extension to support quantum distributed simulation in GPUs using the VirD-GM environment consist in a solution to reduce simulation time.

Regarding simulations with different types of memories, the results showed that it is possible to reduce the simulation time for memories with zero values, and when the number of non-zero values is small, simulation time can be drastically reduced.

Considering distributed simulations, the results showed that it is possible to obtain performance gain with increasing number of clients, except for extreme cases where the number of non-zero values in memory is very small. And transformations with different configurations, but using the same number of clients, have a similar simulation time, this means that the programmer is free to define the configuration of the MPPs that looks better according to the resources available, without worrying about having a great loss of performance.

In future works, we intend to develop strategies that take into account both the hardware architectures and the mathematical characteristics of quantum systems in order to further improve performance and the number of qubits that may be simulated in classical computers.

We also intend to test new approaches that may harness the potential of cloud computing for simulation of quantum algorithms, while the extra communication costs are considered.

Acknowledgements
This work is supported by the Brazilian funding agencies CAPES, FAPERGS (PqG 002/2014, process number 309533/2013-9), MCTI/CNPQ (Universal and PQ, process numbers 448766/2014-0 and 309533/2013-9) and by the CNPq/PRONEX/FAPERGS Green Grid project.

References

[1] A. Avila, A. Maron, R. Reiser, and M. Pilla, "Extending the VirD-GM environment for the distributed execution of quantum processes," in *Proc. of the XIII Symposium on Computational Systems – Cientific Initiation Workshop (WSCAD-WIC)*. Petrópolis: IEEE CPS, 2012, pp. 1–4.

[2] A. d. Avila, A. Maron, R. H. S. Reiser, M. Pilla, and A. Yamin, "GPU-aware distributed quantum simulation," in *Symposium on Applied Computing*. Gyeongju: Proc. of the 29th ACM Symposium on Applied Computing (SAC), March 2014, pp. 860–865.

[3] N. Dickson and M. Amin, "Algorithmic approach to adiabatic quantum optimization," *Physical Review*, vol. 85, p. 032303, Mar 2012.

[4] N. Dickson, M. Johnson, M. Amin, R. Harris, F. Altomare, A. Berkley, P. Bunyk, J. Cai, E. Chapple, and P. Chavez, "Thermally assisted quantum annealing of a 16-qubit problem," *Nature communications*, vol. 4, p. 1903, 2013.

[5] E. Gutierrez, S. Romero, M. Trenas, and E. Zapata, "Quantum computer simulation using the cuda programming model," *Computer Physics Communications*, pp. 283–300, 2010.

[6] M. Henkel, "Quantum computer simulation: New world record on JUGENE," 2010. [Online]. Available: http://www.hpcwire.com/hpcwire/2010-06-28/quantum_computer_simulation_new_world_record_on_jugene.html

[7] M. Hirvensalo, *Quantum computing*, ser. Natural Computing Series. New York: Springer, 2004. [Online]. Available: http://opac.inria.fr/record=b1117053

[8] S. Imre and F. Balazs, *Quantum Computing and Communications - an Engineering Approach*. New Jersey: John Wiley and Sons, 2005.

[9] A. K. Maron, R. H. S. Reiser, and M. L. Pilla, "Correlations from conjugate and dual intuitionistic fuzzy triangular norms and conorms," in *CCGRID 2013 IEEE/ACM International Symposium on Cluster, Cloud and Grid Computing*. NY: IEEE, May 2013, pp. 1–8.

[10] A. Maron, R. H. S. Reiser, M. Pilla, and A. Yamin, "Expanding the VPEÂqGM environment towards a parallel quantum simulation of quantum processes using GPUs," *CLEI Electronic Journal*, vol. 16, no. 3, pp. 1–18, 2013. [Online]. Available: http://www.clei.org/cleiej/papers/v16i3p3.pdf

[11] ——, "Quantum processes: A novel optimization for quantum simulation," *TEMA: Trends in Applied and Computational Mathematics*, vol. 14, no. 3, 2013.

[12] M. A. Nielsen and I. L. Chuang, *Quantum Computation and Quantum Information*. Cambridge University Press, 2000.

[13] K. D. Raedt, K. Michielsen, H. D. Raedt, B. Trieu, G. Arnold, M. Richter, T. Lippert, H. Watanabe, and N. Ito, "Massive parallel quantum computer simulator," http://arxiv.org/abs/quant-ph/0608239, 2006.

[14] E. Rieffel and W. Polak, "An introduction to quantum computing for non-physicists," *ACM Comput. Surv.*, vol. 32, no. 3, pp. 300–335, Sep. 2000. [Online]. Available: http://doi.acm.org/10.1145/367701.367709

[15] M. F. Schmalfuss, A. K. Maron, R. H. S. Reiser, and M. L. Pilla, "qGM analyzer: Suporting of quantum computations in the D-GM environment using c++," in *Proc. of the XIII Symposium on Computational Systems – Cientific Initiation Workshop (WSCAD-WIC)*. Petrópolis: IEEE, 2012, pp. 1–4.

[16] Y. Yonghong, M. Grossman, and V. Sarkar, "Jcuda: A programmer-friendly interface for accelerating java programs with cuda," in *Euro-Par 2009*, Delft,Netherlands, 2009, pp. 1–13.

Journal of Physics: Conference Series **649** (2015) 012005 doi:10.1088/1742-6596/649/1/012005

Middleware for Processing Message Queues with Elasticity Support and Sequential Integrity of Asynchronous Message Processing

Eduardo Henrique Teixeira

Department of Computer Science - CIC, University of Brasilia (UnB), Brasília, DF - Brazil

E-mail: edu.henr@gmail.com

Aletéia Patrícia Favacho de Araújo

Department of Computer Science - CIC, University of Brasilia (UnB), Brasília, DF - Brazil

E-mail: aleteia@cic.unb.br

Abstract. Elasticity in computing refers to dynamically adjusting the amount of allocated resources to process a distributed application. In order to achieve this, mechanisms are needed to avoid the phenomenon of the elasticity threshold detection moving constantly up or down. The existing work fails to deliver sequential integrity of asynchronous messages processing and the asymmetries of data distribution to achieve parallel consumption. This paper fills this gaps and proposes a middleware solution to dynamically analyze the flow of message queue, and a mechanism to increase the parallelized consumption based on the output behavior. An architecture for IOD (Increase On Demand) middleware is presented, with support for the increase and decrease of thread's to cope with the growth of message queues, using the technique of limit-based heuristics over a given period of time and grouping messages into sub-queues based on classification criteria.

1. Introduction

Elasticity is the characteristic of an environment that defines the degree to which a system is able to adapt dynamically to changes in workload, through automatic provisioning and release of resources [10]. In short, elastic computing is the dynamic provisioning of resources [17].

Elastic computing brings significant benefits to application providers, including cost savings and prevention of super- and sub-provisioning of IT resources. This occurs through the monitoring of demand and the acquisition of resources required by applications to achieve a high level of quality [12].

The aim of elasticity is that the amount of resources allocated to a service is what it really needs. It is therefore possible, for example, to reduce the number of servers required to handle the FIFO (First In First Out) [7] message queues (interprocess communication mechanisms to access data asynchronously from multiple processes) of a distributed system and, consequently, to save computational resources. It is also possible to determine the current processing load on a cluster, which can be measured in terms of the flow of outgoing messages from a queue and may be limited by CPU consumption, in order to avoid saturating the servers. This is where a middleware solution can be used: one that is capable of transparently and dynamically implementing the concept of elasticity in order to ensure that processing can adapt to the growth of message queues and, to avoid messages accumulating before they are processed.

In addition to elasticity, fault tolerance is essential in distributed platforms in order to allow one or more processes to fail without harming the rest of the system [20]. In distributed systems, failures are common and may occur due to errors in hardware or software [28]. Hardware failures generally result from physical degradation of components, while software ones occur due to errors in design or implementation, and are also known as bugs [28]. In this context, the tolerance of software failures guarantees correct application behavior, regardless of the number and type of failures occurring [4].

This article proposes the IOD (Increase On Demand) middleware as a valuable tool to ensure elasticity and software fault tolerance in distributed systems with message queue-based architectures, such as high performance and high availability clusters. IOD middleware dynamically analyzes the behavior of the message queue outputs to determine whether to increase or decrease the number of threads responsible for handling messages. The goal of the proposed middleware is to prevent the accumulation of messages to be processed. The IOD also analyzes the behavior of CPU consumption to determine the limits of scalability of servers so as to avoid saturation of the cluster. In addition, it uses mechanisms for fast recovery in case of failures, ensuring that services maintained by the cluster become available again quickly.

IOD middleware's focus is to use elastic computing to determine how to most efficiently utilize the processing cores available in the cluster. To achieve this, the number of threads required to handle the message queues is calculated dynamically according to the behavior of the average queue output and the average CPU consumption of each node. Distributed applications that use message queues as an IPC (Interprocess Communication) mechanism [7] can also benefit from the techniques outlined in this paper. This is possible because the allocation of resources in a cluster is performed at each node, avoiding CPU idle periods.

The remainder of this paper is divided into four sections. Section 2 contains an overview of related elasticity and fault tolerance work. Section 3 describes the architecture of the proposed IOD middleware. Section 4 presents the evaluation and analysis of tests done with the IOD middleware, initially in a simulated environment and then in a real distributed application that uses message queues. Finally, Section 5 presents some conclusions and outlines the next steps of this work.

2. Related Work

To achieve transparent and automatic elasticity, the approaches [14][23] are appropriate when sequential integrity of the processing of messages in the queues is not a functional requirement. The messages in the queues can therefore be redistributed according to the required output, without worrying about the order in which they are processed. However, this can be a disadvantage when it is necessary to maintain sequential integrity in processing the queues for the consumers. In this case, the events generated by consumers need to respect the time constraint in which the messages were generated.

The assignments made by publishers to subscribers can generate asymmetric distribution of data [6][13][23]. These asymmetries are also common in systems with multiple message queues and multiple consumer [13][23]. Identifying the best candidate for consumption of the queues is necessary to reduce latency and avoid saturation of overloaded consumers. Thus, it is possible to determine these candidates identifying the highest average flow [13][23]. The lowest average CPU consumption can also be used as a metric to identify the best candidates for the consumption of new items [13], or by load balancing policy based on flow control [24].

In distributed systems with CPU-bound characteristics, it is possible to explore the transparency of elasticity of an application [11][14][23]. This transparency can be achieved using the metric of load in terms of CPU consumption [11][13][19]. However, if the number of messages remains close to the detection limit of elasticity, problems of rapidly expanding and reducing the thread pool can occur. To solve this problem and keep the output at acceptable levels, [11][13][23] propose the use of limit-based heuristics, for a given amount of time. In this way, the created threads are kept for a longer period, thereby minimizing the cost of restarting them and reducing the latency of messages consumption. This approach is a potential solution to the problem of up-and-down elasticity threshold detection.

Approaches [8][11][12][14][19][23][24][25] mention that elasticity must be achieved transparently and automatically. They also discuss the use of the publisher/subscriber design pattern [6][13][23][24] to achieve elasticity and the decoupling of producers and consumers. With the approaches cited in [11][12][13][19], elasticity is achieved according to the required load demand, which is measured by CPU consumption. Elastic operations can be made transparently without any knowledge of the application, such as the expected time for consuming messages [25]. However, handling elasticity at application level in systems that are composed of redundant components that recover from failures controlled by a middleware also needs to be considered [26].

Many of the approaches to fault tolerance use replication so that another set of processes can take over in case of hardware or software failure [1][2][9][16]. However, techniques such as fast failure recovery [2][3][9][27] are interesting because they avoid the cost of additional memory and communication between replicated processes and the exchange of state information. Papers [3][16][24] propose support for fault tolerance transparently in the middleware. This approach is attractive because it abstracts away from the user application the details of detection and recovery of implemented features. In order to achieve this transparency, [2][16] detect failure by monitoring the connection between processes. A fail-over mechanism that resend a given message transparently to another queue if its initial destination is unavailable [24].

3. Proposed IOD Middleware

To create efficient elastic environments, existing services must be extended with elastic computing capabilities and policies of resource provisioning on demand [15]. The IOD middleware proposes support for elasticity by means of dynamic adaptation of the number of threads for handling messages, based on an analysis of the inbound and outbound throughput of the queues and server CPU consumption. Support for fault tolerance, in IOD middleware, is achieved by detecting the failure and recovering with a quick restart. In this way, the messages in the queues continue to be processed from the point where the failure occurred, without the need for a replication process. These features will be presented in the following sections.

3.1. Message's Handling Architecture

To achieve parallel consumption, it is first necessary to separate the messages from a queue into distinct groups, according to a classification criterion. Each queue thus gives rise to several sub-queues, each indexed by a key that identifies the group. This should be done to increase the number of worker threads that will be responsible for processing the messages, distributing the groups among the various initiated worker threads and thus increasing the throughput by means of parallel consumption.

A system with multiple consumer queues can generate distribution asymmetries [6][13][23], which leads to an imbalance in the system when consuming the messages in the sub-queues. The IOD middleware deals with this by sorting the groups, when the worker threads are created or destroyed, starting from the thread that has the most messages through to the one that has the fewest. A round robin algorithm is then used to distribute these groups among the available worker threads. Furthermore, at each round of consumption, one message from each group is processed in accordance with FQ (Fair Queueing), an algorithm that allows multiple message queues to share the same processing capacity, ensuring fairness in consumption and avoiding starvation due to heavy flows. First Come First Served algorithm [22] is also used to preserve the sequential integrity in the consumption of messages in the same group.

In order to achieve elasticity in the IOD middleware, it is necessary to analyze message queue throughput, the limitation of creating worker threads per CPU and the standard deviation of the average output per group, as well as failure recovery with elasticity, which will be described in the following subsections.

3.2. Analysis of Average Input and Output Flow

During a burst of incoming messages, if a distributed application is unable to process them as quickly as they arrive, an accumulation in the queues will take place, resulting in delays in processing and, consequently, a reduction in the quality of the provided services. To determine the number of worker threads required to cope with queue buildup in the presence of message bursts, it is necessary to calculate the average input and output flow, and the average ratio of incoming to outgoing messages. These calculations are performed from the time queue growth is detected, here called Growth Detection (*GD*), until the Critical Point (CP), which is the moment when new worker threads are created. The *GD* occurs when the ratio of incoming messages to output messages becomes larger than one, causing queueing. The *GD* is detected by Equation 1,

$$\left(\frac{Input}{Output} > 1\right) \tag{1}$$

where *Input* is the number of incoming messages and *Output* is the number of outgoing messages. The mean ratio between the incoming and outgoing messages is called *avgIO*.

New worker threads are created at the time when the accumulation in the queue cannot be dealt with before the end of the maximum time set for handling messages. This moment occurs at the *CP* and the action of increasing worker threads is called *Scale Up* (*SU*). The *CP* is found when the number of messages in the queue (*QueueSize*) is greater than the *MaxQueueSize*, given by Equation 2,

$$(avgTOUT \times (schTIME - (hNOW - hSTART))) \tag{2}$$

where variable *avgTOUT* is the average output flow, *schTIME* is the maximum time for the treatment of messages in the queue, *hNOW* is the current time and *hSTART* is the start time, i.e. the moment at which the value given by Equation 1 becomes true.

The *Scale Down* (*SD*) occurs at the *Exit Point*, which is the time where the worker threads are removed. It is obtained by means of Equation 3,

$$(QueueSize < avgTIN) \tag{3}$$

where *QueueSize* is the current size of the queue, and *avgTIN* is average input flow since the *CP*.

In addition to the *Contention Threads* (*CT*), which are the worker threads intended for burst containment, another set of worker threads is needed, called *Zero Threads* (*ZT*), to consume the messages that have accumulated in the queue between the *GD* and the *CP*. Therefore, various worker threads are created, of types *ZT* or *CT*, and each one is associated with several message sub-queues separated by groups, so that throughput increases as a result of the parallelized consumption. The *CT* and *ZT* threads are shown in Figure 1.

The calculation of the total number of worker threads started in a *SU*, is shown in Algorithm 1. When *QueueSize* is greater than the *CP*, line 2, the value of *avgIO* is used to find the number of *CT* worker threads, line 3. The variable *avgTOUT* divided by the total number of worker threads results in the average flow by thread, line 4, which is used as the rate of consumption,

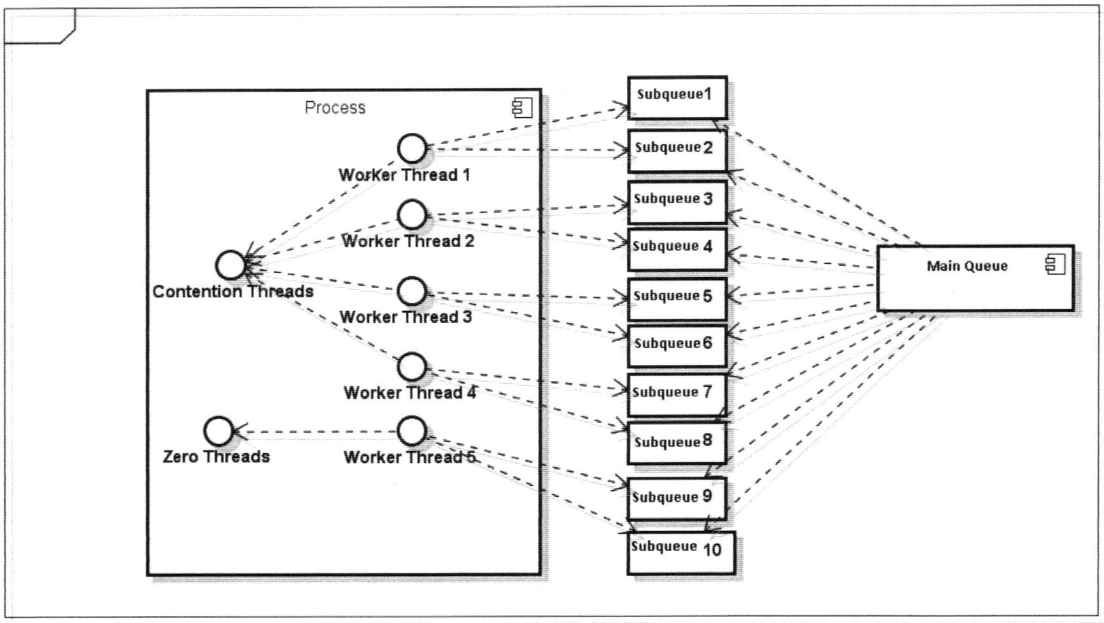

Figure 1: Worker Threads Type *ZT* and *CT* Associated Sub-queues.

line 5, to obtain the number of *ZT* worker threads, line 6, necessary to reset the messages that have been accumulated. To avoid the problem of elasticity threshold detection [11], the IOD middleware starts one worker thread more than the sum of *CT* and *ZT*, line 7. This started thread is a *CT* type and is intended to avoid the problem of elasticity threshold detection [11].

Algorithm 1: Worker Threads Started in the *SU*.

1 MaxQueueSize = (avgTOUT) * (schTIME - (hNOW - hSTART));
2 **if** *(QueueSize > MaxQueueSize)* **then**
3 ContentionThreads = avgIO;
4 ThroughPutByWorkerThread = (avgTOUT / TotalWorkerThreads);
5 ConsumptionRate = QueueSize / (schTIME - (now - hSTART));
6 ZeroThreads = ConsumptionRate / ThroughPutByWorkerThread;
7 TotalThreadsToStart = ContentionThreads + ZeroThreads + 1;
8 **end if**

The calculation of the number of threads that are removed in the *SD*, is shown in Algorithm 2. To perform the *SD* it doesn't matter if the threads that will be removed are the type *CT* or *ZT*, and it is necessary that the value given by Equation 3 is true, as shown in line 1 of Algorithm 2. Then the throughput per thread is calculated, line 2, which is used in the calculation of threads to be removed, line 3. Again, to avoid the phenomenon of elasticity threshold detection [11], one fewer thread is removed, line 4.

3.3. Analysis of CPU Consumption

Knowing the average CPU consumption by worker thread is useful to determine limits for the *SU* (*Scale Up*). Thus, when bursts of incoming messages are detected, a process is started for

Algorithm 2: Worker Threads Removed in the *SD*.

```
if ( QueueSize < avgTIN ) then
    ThroughPutByThread = (avgTOUT / TotalWorkerThreads);
    ThreadsToRemove = TotalWorkerThreads - (avgTIN / ThroughPutByThread);
    ThreadsToRemove = ThreadsToRemove - 1;
end if
```

measuring average CPU consumption, per worker thread and per cluster node, to determine if the required throughput to contain growth and drain the queue can be achieved without saturating the server CPU.

3.4. Analysis of Standard Deviation of Average Output Flow per Group

The bursts of incoming messages of a distributed system that handle message queues may originate from several different groups. To handle the bursts from n groups, the IOD middleware creates a number of worker threads, according to the average output flow, calculated during the growth of the queue, as described in the analysis of the average input and output flow, presented in Section 3.2.

However, when the bursts originate from a small group of messages, creating many worker threads can mean a waste of resources and, depending on the consumption throughput, may not be enough to contain the growth of the queue and still generate recursive *SUs* (*Scale Up*), without solving the problem of queue growth. To solve this problem, before the *SU* and during the burst of messages, the average output flow per group is measured and, this will be used to calculate the standard deviation at the moment the *SU* is performed.

A random variable in a normal distribution has a 95% chance of being less than two standard deviations from its mean [5]. Thus, since the average output flow was normalized, in at most one message from each group per round of the FQ algorithm, after n rounds, when the average output moves more than two standard deviations away from the mean, it is possible to detect bursts coming from specific groups. In this way, the amount of worker threads created is limited by the number of groups having an average output flow greater than twice the standard deviation identified during the burst. This approach aims to create worker threads specifically for the groups in which bursts where detected.

3.5. Fault Tolerance

In the IOD middleware, software fault tolerance features were supported for process failure, but hardware failures were not considered. Therefore, the messages in the queues could continue to be processed from the point where the failure occurred, without requiring the use of replication processes.

To support fast failure recovery mechanisms [2][3][9][27], if a process fails and the number of messages in the queue is greater than the *CP* (*Critical Point*), mechanisms of elasticity in the middleware are required to perform the *SU* (*Scale Up*). Thus, when the processes are restarted after a failure, if the input bursts continue, the *SU* occurs rapidly by detecting the *CP* larger than the *QueueSize*. The incomplete message being processed by the failing process goes back to the queue to be handled by the process when it restarts. However, at this time another mechanism is needed to provide elasticity, if the bursts ended after the failure point but before the process restart. In this situation, if the number of messages in the queue is above *avgTOUT* x *schTIME* (the measurements of throughput) before deciding to perform the *SU*, it's performed using only 10% of *schTIME*. This is necessary to perform the *SU* more quickly, which is the case when the bursts end after the failure and before the quick restart of the process. The same scenario occurs when bursts of messages do not stop, even after the quick restart of the process.

The efficient implementation of fault tolerance in a distributed environment, relies on a mechanism of elasticity, exactly as implemented by the IOD middleware.

4. Results

The tests for the IOD middleware were performed in both a simulated environment and a real environment. For the simulated tests of the IOD middleware, a simulator was created with two threads: a producer that writes packets to the message queue and a consumer that reads and delivers the messages to the IOD middleware. For the simulated tests, *schTIME* was given the value of 80 seconds because this is maximum time of message handling required by the real distributed application that will use the IOD middleware. The production thread was programmed to generate messages at a rate five times greater than the consuming thread. The purpose of this setting is to cause message accumulation, and demonstrate the operation of the elasticity proposed by the IOD middleware. After testing in a simulated environment, the IOD middleware was deployed in a real distributed application that uses message queues to validate if the middleware achieved the goal of dynamically adapting to the throughput conditions and avoided queueing. The tests were divided into five groups, as described below.

4.1. Creating and Terminating Threads Dynamically

The tests presented in this section are intended to verify the functioning of dynamic creation and termination of worker threads, based on the average flow of incoming and outgoing messages.

Tests were carried out with the producer creating a burst of packets every 10 milliseconds and the consumer processing them every 50 milliseconds, i.e. generation of messages at a rate five times greater than consumption. Accordingly, the experiments showed that there was an adequate increase of the number of worker threads to increase the flow rate of consumption, as shown in Figure 2, in which the x-axis corresponds to elapsed time and the y-axis corresponds to the value of the variables *QueueSize*, *MaxQueueSize*, *avgTIN* and *avgTOUT*. These variables correspond respectively to the values of the queue length; the value given by Equation 2, average input flow, and the average output flow.

In the scenario presented in Figure 2, the *CP* was detected with 1130 messages in the queue, 14 seconds after the start of the message burst, as shown in Table 1. At 9 seconds after the *SU*, the middleware decided to terminate five threads so that the average output flow would be compatible with the average input flow. Here, the IOD middleware proved satisfactory in terms of creation and termination of worker threads based on the throughput, because it was able to create 12 worker threads and end the bottleneck in a period of 23 seconds, i.e., within the limit of message handling defined as 80 seconds. The number of messages reached a controlled level, because the stabilization of input and output messages occurred with an acceptable number (33) in which the number of messages in the queue doesn't grow beyond the *Critical Point*.

Table 1: Times of the Elasticity Algorithm.

Moment	Duration (sec.)	Input Messages	Output Messsges	Threads	Queueing
Start	14	99	19	1	0
Scale Up	9	99	197	12	1.130
Scale Down	2	99	99	7	33

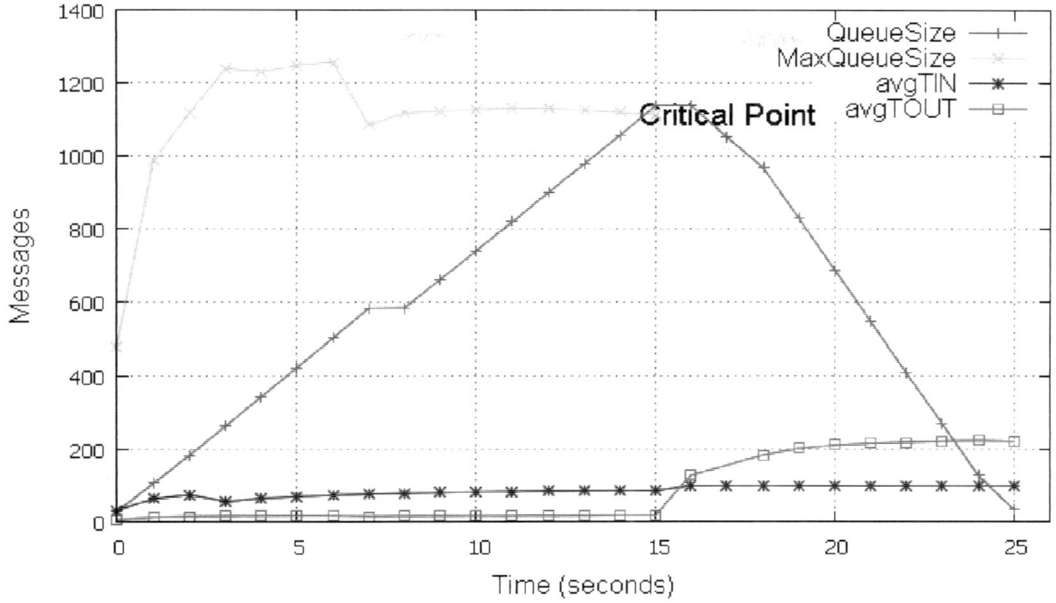

Figure 2: Message Queue with Creation and Removal of Worker Threads.

4.2. Creation of Threads Limited by Average CPU Consumption

The purpose of this test is to determine if the elasticity algorithm prevents saturation of CPU consumption on the servers by checking the behavior of the increase of worker threads limited by the average CPU consumption. The worker threads were configured to use the busywait technique in order to simulate high CPU usage in the simulation. The tests were performed with the producer thread generating messages every 10 milliseconds, and the consumer thread processing them every 50 milliseconds, resulting in message accumulation.

The results of the message queue, CPU and threads are shown in Figures 3a and 3b. In Figure 3a, the x-axis is the elapsed time, and the y-axis shows the variables *QueueSize*, *MaxQueueSize*, the *avgTIN* and *avgTOUT*. These variables correspond, respectively, to the values of the queue length, the value given by the Equation 2, average input flow, and the average output flow. In Figure 3b, the x-axis is the elapsed time in seconds and the y-axis is the percentage of CPU usage of the process and the machine, or the number of threads in the process.

As can be seen in Figure 3, the reduction in the number of messages occurs more slowly because the number of threads is limited based on the average CPU usage of the worker threads so as not to exceed 100% of the server. Like the previous experiment the decrease is also linear but it happens in a lower rate, as shown in Figure 2, because not all the worker threads necessary for processing the entire queue in the required time were created. According to Table 2, the middleware should have created five worker threads based on the average input flow. However, as shown in Table 3, the average CPU consumption by worker thread was 15.28%, and would exceed the limit of the residual server CPU, which was 83.60%. The middleware therefore created only four worker threads (1 fewer) to avoid saturating the CPU usage of the machine. At the time of *Scale Down*, the average output flow was at 107 messages per second and therefore below the average input flow, which was 115.5 messages per second. As a result, the middleware did not detect the need to remove worker threads.

It is important to consider the saturation of the environment in order to make the distributed

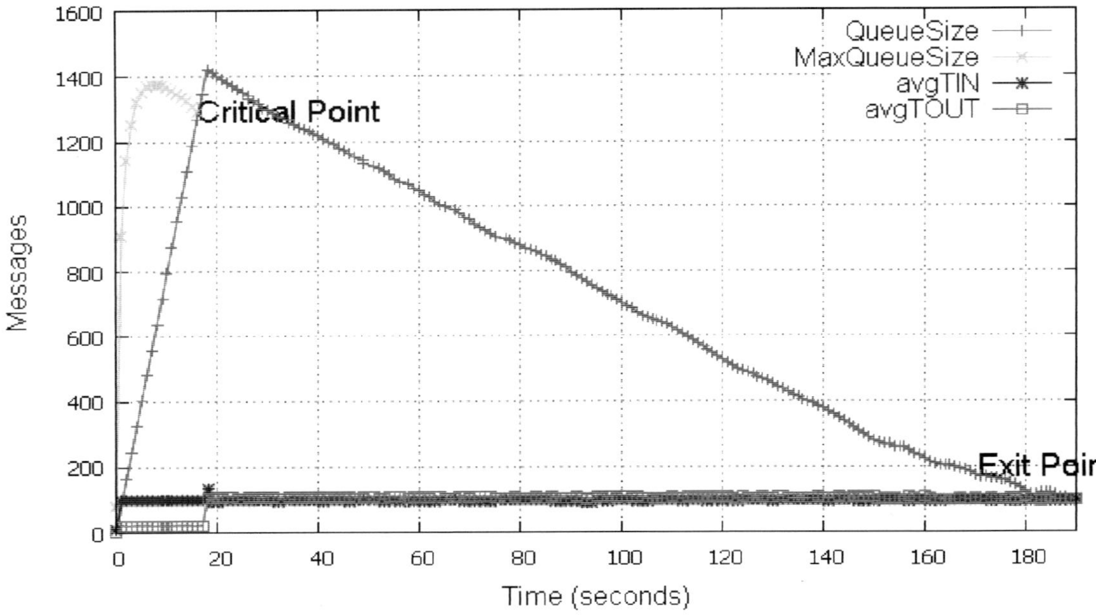

(a) Message Queue.

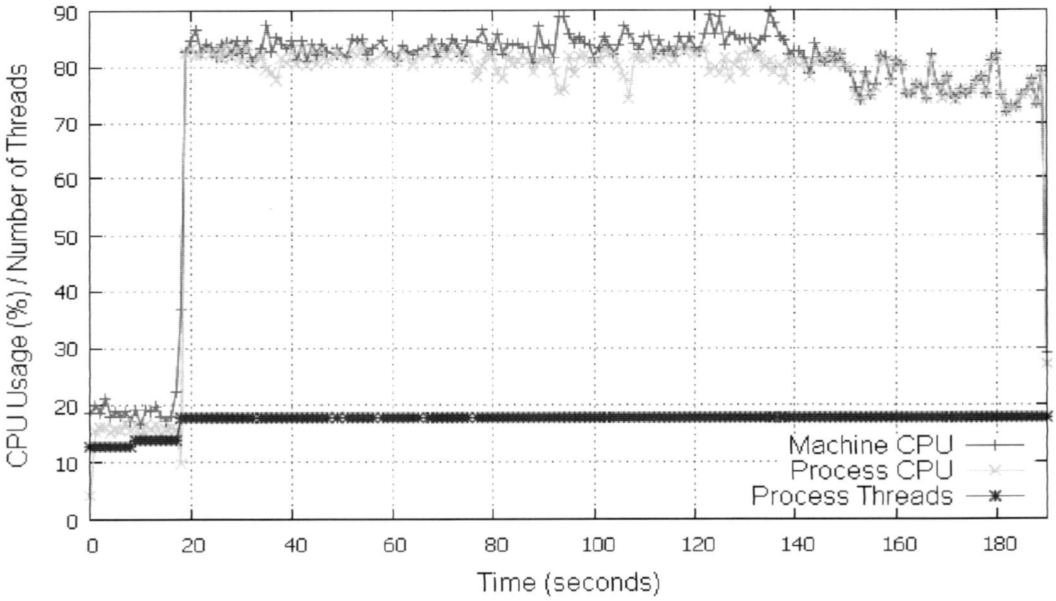

(b) CPU and Threads.

Figure 3: Behavior of Message Queue, CPU and Threads with CPU Limits.

application capable of dynamically identifying the limits of the workload that it is capable of processing given the size of its infrastructure. If the system identifies that it needs to increase its capacity but cannot do so due to imposed limitation, it can generate alarms so that additional adjustments can be made according to the required load.

Table 2: Behavior of the Threads with CPU Limits.

Moment	Threads Needed	Created Threads	Removed Threads
Scale Up	5	4	N/A
Scale Down	N/A	N/A	0

Table 3: Behavior of the CPU in the Elasticity Algorithm with Limits.

Moment	Worker Thread Mean	Total CPU Needed	Residual CPU
Scale Up	15.28%	91.72%	83.60%
Scale Down	N/A	N/A	N/A

4.3. Creating Threads Limited by Standard Deviation

These tests aim to verify the operation of creating worker threads limited by the number of groups, whose average output flow exceed two standard deviations from the mean. The idea is to analyze if, during a burst of packets from a few groups, the worker threads are created based only on the analysis of the average output flow, because in this case resources are wasted with threads that will not contain the growth of the queue. This procedure is designed to test the behavior of the IOD middleware before the detection of groups that have isolated bursts of messages and check whether the decreased use of resources occurred through the creation of fewer worker threads.

Consecutive burst tests were performed for three different groups, and once again the producer thread generated messages at intervals of 10 milliseconds and the consumer thread processed at intervals of 50 milliseconds. For the first group the bursts were generated for the first 16 seconds, for the second group they were generated from seconds 17-38, and for third group they were generated from seconds 39-75. Figure 4 shows the behavior of the message queue, where the x-axis corresponds to the elapsed time and the y-axis corresponds to the value of the variables *QueueSize*, *MaxQueueSize*, *avgTIN*, *avgTOUT* and *Threads*. These variables correspond, respectively, to the values of the queue length, the value given by Equation 2, the average input flow, the average output flow, and the number of threads in the process (on the scale of 100:1).

Despite three threads having been created, one for each group at the three critical points, the message queue cannot be contained, as shown in Figure 4. This behavior occurs because the message generation is five times higher than consumption, and the middleware cannot create more than one thread per group due to the restriction of sequential integrity in processing data from the same group. Therefore, despite the increase in the number of threads, the number of messages in the queues does not decrease because there is no possibility of separating the data from the same group for parallel processing. Moreover, it is possible to observe in Figure 4 that, after the first *SU*, despite the queue size being above the value of *MaxQueueSize*, the next *SU* only occurs after the queue doubles in size compared to the previous *SU*, to avoid *SUs* occurring very close to each other. It was possible to save resources by creating worker threads for the treatment of specific groups with isolated bursts and, hence, fewer threads than determined by the algorithm associated with the analysis of the flow, as shown in Table 4.

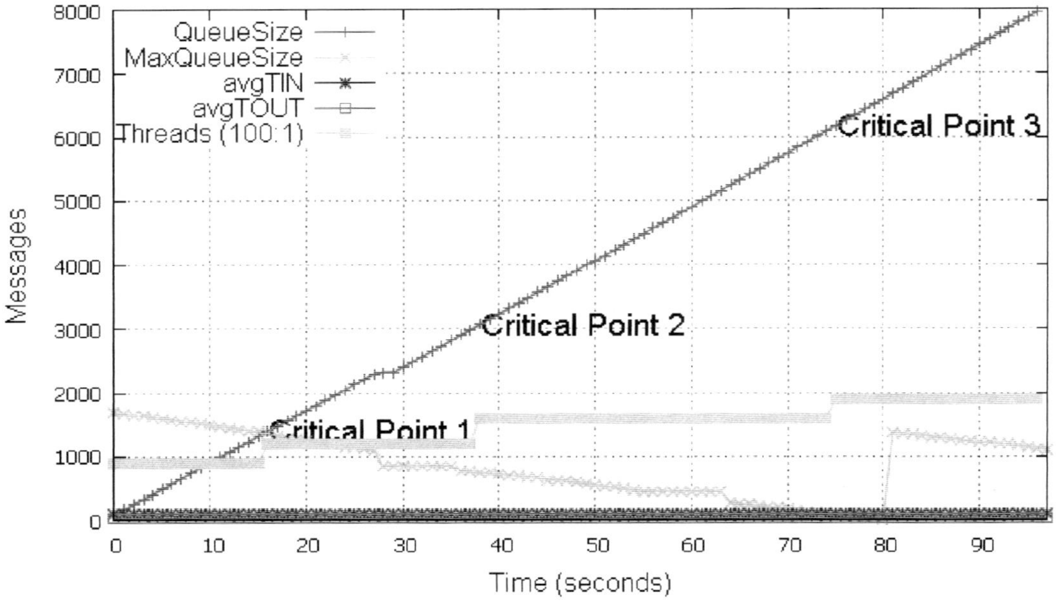

Figure 4: Behavior of the Message Queue with Bursts in Three Groups.

Table 4: Creating Threads Limited by Standard Deviation.

Moment	Threads (Throughput)	Threads (Standard Deviation)
Critical Point 1	6	1
Critical Point 2	9	1
Critical Point 3	11	1

4.4. Elasticity with Fault Tolerance

The objective of this test is to validate if, after a failure while processing messages, the elasticity is initiated without the presence of packet bursts, and after a quick restart of the process. The simulation involved a process crash followed by a quick restart, without any message bursts and the queue containing more messages than *MaxQueueSize*.

In Figure 5, the x-axis corresponds to the elapsed time and the y-axis corresponds to the value of the variables *QueueSize*, *MaxQueueSize*, *avgTIN* and *avgTOUT*. These variables correspond, respectively, to the values of the queue length, the value given by Equation 2, the average input flow, and the average output flow. After restarting the process, despite the *QueueSize* exceeding the *MaxQueueSize*, an increase in the number of consumer threads only occurs after 10% of *schTIME* (8 seconds). This approach aims to analyze the behavior of input and output flows before a *SU*, generating the averages for the calculations of the number of worker threads required to drain the message queue before the end of the maximum message processing time (*schTIME*). A few seconds after the restart of the process, increases in the consumption of messages indicate an increase in the number of worker threads to drain the messages from the queue before the end of *schTIME*. This behavior demonstrates that, even with the process crash, the IOD middleware can be elastic and tolerant against a demand generated and terminated before the quick restart.

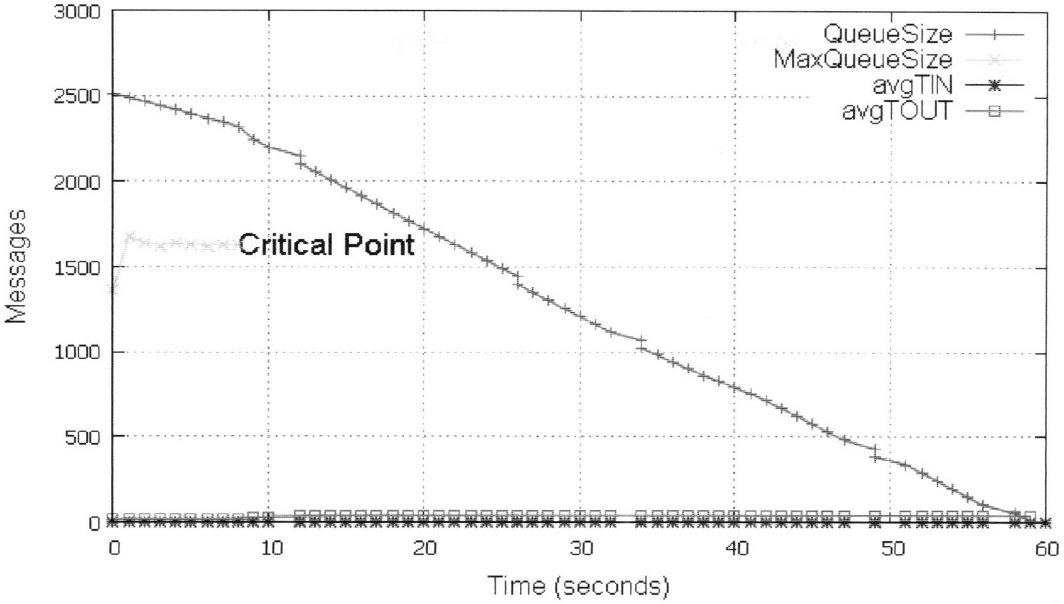

Figure 5: Behavior of Message Queue after the Quick Restart.

4.5. Distributed Real Application

This section analyzes the tests of the IOD middleware in a real distributed application running on a high performance cluster that consists of 11 servers, each one with four cores and four gigabytes of memory. All 11 servers run Microsoft Windows 2008 R2 64 bit operating system, and are virtualized from two servers 360C BLADE G8, each one with two Intel Xeon Quad Core/Eight Threads model E5540 2.53 GHz processors and 96 gigabytes of memory.

Each server in the cluster runs a distributed application with 19 processes, one being the coordinator that performs IPC between coordinated processes through MSMQ (Microsoft Message Queue). Each of these processes is responsible for a distinct feature of the distributed application and has an independent message queue which is fed every 80 seconds. Before using the IOD middleware in a real distributed application, it was necessary to modify the IPC protocol by adding a group identifier to indicate with which group the messages are associated. This procedure allows the IOD middleware to group related messages and to then divide sub-queues between worker threads.

Initially, data was collected for the time required for consumption of the messages from the queues of the 11 servers, over a period of 6 hours, under heavy message traffic. It was found that the average time for processing messages during this period was 5.2 seconds. Given that the maximum treatment time of the messages ($schTIME$) is 80 seconds, the distributed application's servers used the CPU in very short bursts and, were idle most of the time (93.5%). In order to utilize the available resources on this platform more efficiently, the 11 servers of the cluster were grouped into a single server. Whereas the treatment time of each message is not affected by grouping servers (because all nodes have the same hardware configuration) the time for processing messages is 6.5% (5.2 seconds of 80) on average. Therefore, to reduce the number of servers by a factor of 11, an increase of 71.5% of the server's processing capacity is required. Doubling the capacity of a server is more than sufficient in this case, so the virtualized server had its configuration increased from four to eight cores and four to eight gigabytes of memory.

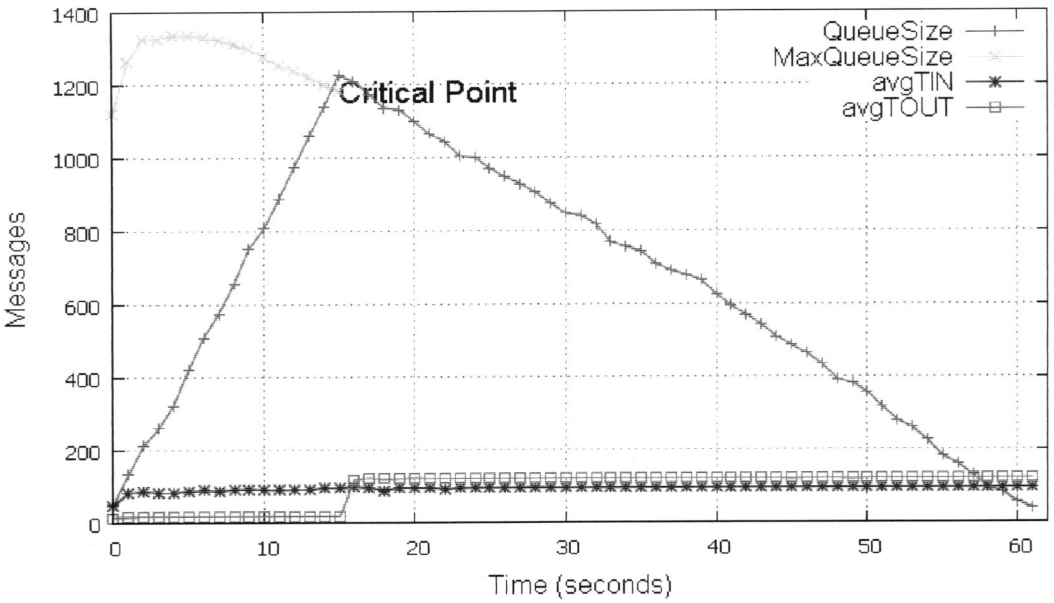

Figure 6: Behavior of Message Queues in a Distributed Real Application.

In this setup, when bursts of messages occur, the distributed application has a buffer of CPU and memory to contain the queueing and ensure scalability of the environment.

After running the real distributed application, it was found that one of the 19 processes had a backlog of messages in its queue, as shown in Figure 6. To contain this accumulation, 15 seconds after the start of the burst of messages the IOD middleware enabled the creation of more worker threads, by observing that the *QueueSize* (1.226) was greater than the *MaxQueueSize* (1182). Six more worker threads were created with an estimated time of 65 seconds to drain the message queue, below the maximum processing time (*schTIME*) of 80 seconds. After the creation of the worker threads the *avgTOUT* was greater than the *avgTIN*, indicating increased consumption parallelism achieved by means of elasticity, to contain the burst of messages as shown in Figure 6.

As the average CPU usage per worker thread was 0.10%, and the residual CPU server was 99.57%, the IOD middleware did not detect the need to limit the creation of worker threads due to CPU consumption. The IOD middleware also detected no cases of groups with isolated bursts, thereby not creating worker threads based on the standard deviation of the average output flow per group.

The tests have shown that is possible to use the IOD middleware to implement elasticity in a real distributed application through the creation and termination of worker threads in accordance with the workload. Another important contribution was to use consumption parallelism in message queues through the clustering of correlated messages in sub-queues. The use of elasticity in the IOD middleware led to a decrease in infrastructure costs, due to being able to reduce the number of servers needed to process the message queue and making more efficient use of each node cluster, to avoid CPU idle time. Creating mechanisms that provide increased throughput of consumption significantly improves the efficiency and the use of infrastructure resources, since the same mass amount of data can be processed in less time and by fewer servers.

5. Conclusions and Future Work

The increasingly high cost of infrastructure equipment shows that it is essential to use available resources optimally and efficiently. Dynamically and automatically scaling resources close to the amount of required processing is, important to service providers, as it impacts directly on the cost savings with no loss of service quality.

The IOD middleware proved that it is possible to achieve the requirements of elasticity, adapting to the conditions of the processing, load by increasing or decreasing the number of worker threads according to demand. Moreover, the IOD middleware proved that it is capable of optimizing the use of IT resources, through task parallelism, using multiple worker threads to process the message queues. This study was able to effectively achieve a saving in infrastructure resources, reducing the number of servers from 11 to a single server with double the current configuration needed to process the message queues of a real distributed application. This reduced the cost of the hardware required to run the distributed application and the redundant equipment can be used for other applications as well as in the failover of the cluster that is hardware available in the case of failure or abnormal termination of the previously active system.

A contribution from this paper was the creation of Equations 2 and 3 to dynamically determine the critical points of entry and exit for the parallel processing of message queues, according to the input and output flow, and the detection of growth given by Equation 1. These equations are based on the dynamic behavior of each server and the time limits imposed for the processing of messages. None of the workers cited in Section 2 has this transparent and dynamic adjusting of resources at middleware level according to the necessary time for consuming messages based on QoS (Quality of Service) parameters, defined with each client of the distributed application that uses message queues, such as maximum time for message's handling or maximum CPU usage. In additional, none of the approaches studied concern themselves with sequential integrity of asynchronous messages processing, nor with the asymmetries of data distribution to achieve parallel consumption, which is achieved by the IOD middleware.

A current limitation of the IOD middleware is linked to the fact that it cannot parallelize the processing of messages in the same group, due to the sequential integrity constraint that imposes ordered processing. As future work, we intend to implement instruction pipelining to optimize the execution of the processing of messages in the same group, without generating sequential integrity conflicts.

References

[1] Heithem Abbes, Christophe Cerin, Mohamed Jemni, and Walid.Saad. Fault tolerance based on the publish-subscribe paradigm for the bonjourgrid middleware. *11th IEEE/ACM International Conference on Grid Computing*, pages 57–64, 2010.

[2] Tekin Bicer, Wei Jiang, and Gagan Agrawal. Supporting fault tolerance in a data-intensive computing middleware. *2010 IEEE International Symposium on Parallel & Distributed Processing (IPDPS)*, pages 1–12, 2010.

[3] Marcela Castro, Dolores Rexachs, and Emilio Luque. Transparent fault tolerance middleware at user level. *2012 International Conference on High Performance Computing and Simulation (HPCS)*, pages 566–572, 2012.

[4] George Coulouris, Jean Dollimore, and Tim Kindberg. *Distributed Systems Concepts and Design*. Publisher Pearson Prentice Hall, fourth edition, 2009.

[5] Douglas Downing and Jeffrey Clark. *Applied Statistic*. Publisher Saraiva, third edition, 2011.

[6] Wenjing Fang, Beihong Jin, Biao Zhang, Yuwei Yang, and Ziyuan Qin. Design and evaluation of a pub/sub service in the cloud. *International Conference on Cloud and Service Computing*, pages 32–39, 2011.

[7] John Shapley Gray. *Interprocess Communications in Linux: The Nooks & Crannies*. Publisher Pearson Prentice Hall, first edition, 2003.

[8] Yike Guo, Rui Hanm, Benjamin Satzger, and Hong-Linh Truong. Programming directives for elastic computing. *Internet Computing and IEEE*, pages 72–77, 2012.

[9] Chen He, Derek Weitzel, David Swanson, and Ying Lu. Hog: Distributed hadoop mapreduce on the grid. *SC Companion: High Performance Computing and Networking Storage and Analysis*, pages 1276–1283, 2012.

[10] Nikolas Roman Herbst, Samuel Kounev, and Ralf Reussner. Elasticity in cloud computing: What it is, and what it is not. In *Proceedings of the 10th International Conference on Autonomic Computing (ICAC 13)*, pages 23–27, San Jose, CA, 2013. USENIX.

[11] Shigeru Imai, Thomas Chestna, and Carlos A. Varela. Elastic scalable cloud computing using application-level migration. *IEEE/ACM Fifth International Conference on Utility and Cloud Computing*, pages 91–98, 2012.

[12] Philipp Leitner, Christian Inzinger, Waldemar Hummer, Benjamin Satzger, and Schahram Dustdar. Application-level performance monitoring of cloud services based on the complex event processing paradigm. *5th IEEE International Conference on Service-Oriented Computing and Applications*, pages 1–8, 2012.

[13] Ming Li, Fan Ye, Minkyong Kim, Han Chen, and Hui Lei. A scalable and elastic publish/subscribe service. *IEEE International Parallel & Distributed Processing Symposium*, pages 1254–1265, 2011.

[14] Ricky K.K. Ma, King Tin Lam, Cho-Li Wang, and Chenggang Zhang. A stack-on-demand execution model for elastic computing. *39th International Conference on Parallel Processing*, pages 208–217, 2010.

[15] Paul Marshall, Henry Tufo, and Kate Keahey. Provisioning policies for elastic computing environments. *26th International Parallel and Distributed Processing Symposium Workshops & PhD Forums*, pages 1085–1094, 2012.

[16] Rolando Martins, Priya Narasimhan, Lu Lopes, and Fernando Silva. Lightweight fault-tolerance for peer-to-peer middleware. *29th IEEE International Symposium on Reliable Distributed Systems*, pages 313–317, 2010.

[17] Julien Perez, Ccile Germain-Renaud, Balzs Kgl, and Charles Loomis. Responsive elastic computing. *ACM/IEEE Conference on International Conference on Autonomic Computing*, pages 55–64, 2009.

[18] Douglas Schmidt, Michael Stal, Hans Rohnert, and Frank Buschman. *Pattern-Oriented Software Architeture.* Publisher John Wiley & Sons, first edition, 2000.

[19] Akiyoshi Sugiki and Kazuhiko Kato. An extensible cloud platform inspired by operating systems. *Fourth IEEE International Conference on Utility and Cloud Computing*, pages 306–311, 2011.

[20] Andrew S. TANENBAUM and Maarten Van Steen. *Distributed systems: principles and paradigms.* Publisher Pearson Prentice Hall, second edition, 2008.

[21] Andrew S. Tanenbaum and David J. Wetherall. *Computer Networks.* Publisher Pearson Prentice Hall, fifth edition, 2010.

[22] Andrew S. TANENBAUM and Albert S. Woodhull. *Operating Systems Design and Implementation.* Publisher Pearson Prentice Hall, third edition, 2007.

[23] Nam-Luc Tran, Sabri Skhiri, and Esteban Zimnyi. Eqs: an elastic and scalable message queue for the cloud. *Third IEEE International Conference on Cloud Computing Technology and Science*, pages 391–398, 2011.

[24] Ahmed El Rheddane, Noel de Palma, Alain Tchana and Daniel Hagimont. Elastic Message Queues. *IEEE International Conference on Cloud Computing*, 2014.

[25] Rodrigo da Rosa Righi, Vinicius Facco Rodrigues, Cristiano Andr da Costa, Guilherme Galante, Luis Bona, Tiago Ferreto. AutoElastic: Automatic Resource Elasticity for High Performance Applications in the Cloud. *IEEE Transactions on Cloud Computing*, 2015.

[26] Maria Toeroe, Neha Pawar and Ferhat Khendek. Managing Application Level Elasticity and Availability. *10th CNSM and Workshop*, 2014.

[27] Jun Wang, Yu Deng Jian-Wen Chen, and Di Zheng. Research of the middleware based fault tolerance for the complex distributed simulation applications. *International Conference on Computational Intelligence and Software Engineering (CiSE)*, pages 1–4, 2009.

[28] Jonathan M. Smith. A survey of software fault tolerance techniques. pages 165–170, 1986.

Towards Cloud-based Asynchronous Elasticity for Iterative HPC Applications

Rodrigo da Rosa Righi[1], Vinicius Facco Rodrigues[1], Cristiano André da Costa[1], Diego Kreutz[2] and Hans-Ulrich Heiss[3]

[1] Applied Computing Graduate Program - Unisinos - Av. Unisinos, 950 – São Leopoldo, RS, Brazil
[2] SnT, University of Luxembourg - 4, rue Alphonse Weicker L-2721 Luxembourg
[3] Technische Universität Berlin - Sekretariat EN 6, Einsteinufer 17 D-10587 Berlin

E-mail: rrrighi@unisinos.br, viniciusfacco@live.com, cac@unisinos.br, diego.kreutz@uni.lu, hans-ulrich.heiss@tu-berlin.de

Abstract. Elasticity is one of the key features of cloud computing. It allows applications to dynamically scale computing and storage resources, avoiding over- and under-provisioning. In high performance computing (HPC), initiatives are normally modeled to handle bag-of-tasks or key-value applications through a load balancer and a loosely-coupled set of virtual machine (VM) instances. In the joint-field of Message Passing Interface (MPI) and tightly-coupled HPC applications, we observe the need of rewriting source codes, previous knowledge of the application and/or stop-reconfigure-and-go approaches to address cloud elasticity. Besides, there are problems related to how profit this new feature in the HPC scope, since in MPI 2.0 applications the programmers need to handle communicators by themselves, and a sudden consolidation of a VM, together with a process, can compromise the entire execution. To address these issues, we propose a PaaS-based elasticity model, named AutoElastic. It acts as a middleware that allows iterative HPC applications to take advantage of dynamic resource provisioning of cloud infrastructures without any major modification. AutoElastic provides a new concept denoted here as asynchronous elasticity, i.e., it provides a framework to allow applications to either increase or decrease their computing resources without blocking the current execution. The feasibility of AutoElastic is demonstrated through a prototype that runs a CPU-bound numerical integration application on top of the OpenNebula middleware. The results showed the saving of about 3 min at each scaling out operations, emphasizing the contribution of the new concept on contexts where seconds are precious.

1. Introduction

One of the key features of the cloud includes the elasticity, where users can scale at any moment their resource consumption up or down according to either the demand or the desired response time [1, 2]. Considering the HPC landscape and a very long running parallel application, a user may want to increase the number of instances to try and reduce the completion time of the application. On the other hand, if an application is not scaling in a linear or close to linear way, and if the user is flexible with respect to the completion time, the number of instances can be reduced. This results in a lower nodes $\times$ hours index, and thus in a lower cost and energy saving. Although there are benefits to HPC systems, cloud elasticity has been more extensively explored on client-server Web architectures, such as video on demand, online stores,

BOINC applications, e-governance and Web services [2]. As illustrated in Figure 1, a typical strategy in this context uses horizontal cloud elasticity to replicate virtual machine instances in a datacenter [3, 4]. Despite transparent to the user, this kind of mechanism is suitable on loosely coupled programs in which replicas do not establish communication among themselves [5, 6].

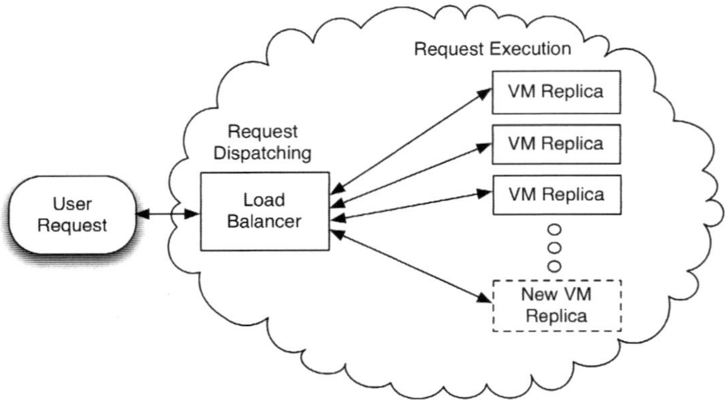

Figure 1. Standard cloud elasticity mechanism: horizontal elasticity and an elasticity controller acting as load balancer.

Although pertinent for bag-of-tasks and key-store HPC applications, replication techniques and centralized load balancers are not useful by default to implement elasticity on tightly-coupled HPC applications, as those modeled as Bulk-Synchronous Parallel (BSP), Divide-and-Conquer or pipeline [2, 7]. This happens because any resource (de)allocation causes a process reorganization as well as the updating of the whole communication topology, not only the interaction between the load balancer and the target replicas. In addition, there is a problem related to virtual machine consolidation, which can result in a sudden termination of a process and its disconnection from the communication topology; and consequently, resulting in the application crash. Most parallel applications have been developed using the MPI 1.x, which means that they do not have any support for changing the number of processes during the execution, so applications cannot explore elasticity without an appropriate support [8]. While this changed with MPI version 2.0, significant effort is needed at application level both to manually change the process group and to redistribute the data to effectively use a different number of processes.

Figure 2 (a) depicts a situation in which elasticity controls are implemented inside the application code using the cloud-supported API. This strategy requires user expertise on cloud monitoring, besides the selection of the appropriate points to insert the calls. Part (b) of Figure 2 explores the use of an elasticity controller outside the application, which is normally offered as optional component in platforms such as Amazon and Windows Azure [9]. Resource monitoring, as well as allocation and deallocation of VMs are tasks belonging to the controller, but users must both insert calls in their applications and handle the communication topology reorganization. The call of the *elasticity()* method represents a link between the application and the controller, so the use of a controller without it has no effect in load balancing because of the application is not able to detect and use new resources [10]. To bypass these limitations, some approaches impose code rewriting [2, 11], previous configuration of elastic rules and actions [2, 12, 13, 14], former knowledge of the application phases [2, 12, 13, 14], and the stop-reconfigure-and-go [2] mechanism, to obtain gains from resource reconfiguration.

Aiming at providing cloud elasticity for HPC applications in an efficient and transparent

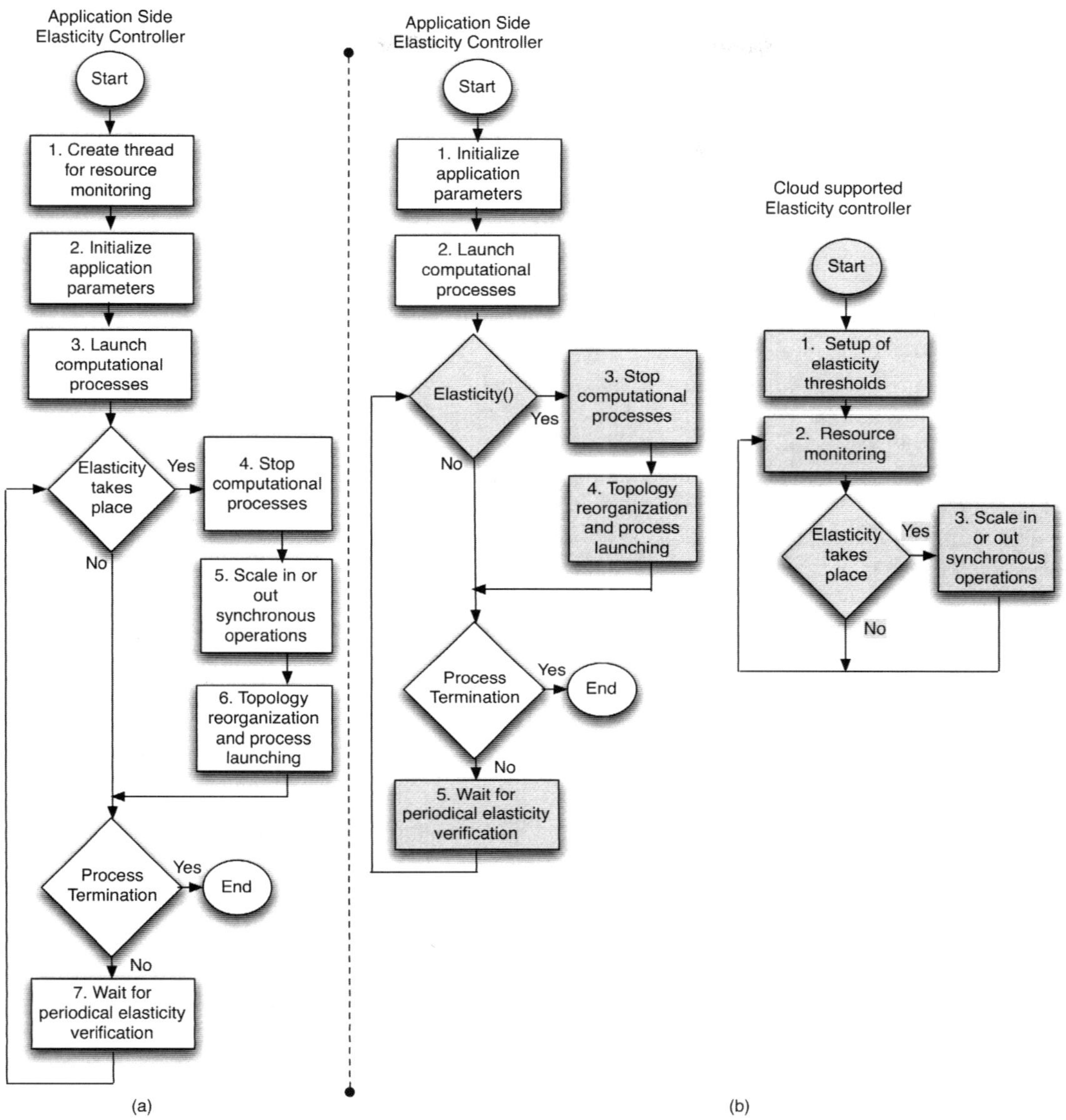

Figure 2. Different approaches for cloud elasticity: (a) elasticity actions are managed directly at application code; (b) use of an elasticity controller outside the application, which can offer a concomitance between elasticity and application's processes actions. However, this approach is not transparent to the users, who need to test elasticity actions and reorganize the communication topology by themselves. On the other hand, AutoElastic offers the concept of asynchronous elasticity by offering a framework to address the aforementioned asynchronism totally transparent from the user. To accomplish this, AutoElastic addresses the shading boxes of part (b) which are put apart from the user responsibility.

manner, we are proposing a PaaS-based model called **AutoElastic**[1]. Particularly, AutoElastic is focusing on master-slave iterative applications, but offers a flexible framework to support

[1] Project website: http://autoelastic.com

other HPC programming styles, such as pipeline and BSP. AutoElastic's contribution relies on the concept of asynchronous elasticity: transparent resource and process reorganization at user perspective, neither blocking nor stopping the application execution at any resource allocation or deallocation action. To accomplish this, AutoElastic provides a framework with a controller that transparently manages horizontal elasticity actions, *i.e.*, without requiring any application modification or adptation. Taking at starting point Figure 2 (b), our approach offers a framework to hide all shading boxes from the user. Although the standard use of a controller enables the setup of VMs in parallel to the application runtime, the benefits of the new resources are not transparent to the users. As discussed earlier, scaling in operations also appear as a problem in the standard utilization of a controller, since a consolidation of one or more VMs will sudden terminate the processes residing on them, which can imply in a premature application ending.

The proposed model assumes that the target HPC application is iterative by nature, *i.e.*, it has a time-step loop. This is a reasonable assumption for the most of MPI programs [15, 16], so this does not limit the applicability of our model. This article describes AutoElastic and a prototype developed with OpenNebula. Tests with a CPU-bound numeric integration application show gains up to 26% when using AutoElastic in comparison with a static provisioning. The remainder of this article will first introduce the related work in Section 2, pointing out open issues and research opportunities. Section 3 is the main part of the article, describing AutoElastic's framework together with asynchronous elasticity concept in details. Section 4 describes a prototype implementation. Evaluation methodology and results are discussed in Sections 5 and 6. Finally, Section 7 emphasizes the scientific contribution of the work and notes several challenges that we can address in the future.

2. Related Work

Elasticity is one of the most attractive features of cloud computing because it allows users to scale resources on-demand. There are different ways of using the elasticity provided by cloud infrastructures, such as manual setup [17, 18, 19], and by pre-configuration of reactive elastic mechanisms [20, 9]. While the former is not suitable for applications that need automatic and transparent elasticity, the latter implies in rather complicated tasks for non-cloud savvy users (e.g., define thresholds and elasticity actions).

Amazon AWS (`http://aws.amazon.com`), Nimbus (`http://www.nimbusproject.org`) and Windows Azure (`http://azure.microsoft.com`) are examples of systems that provide elasticity through pre-configured reactive mechanisms. Middleware solutions for building elastic computing infrastructures, such as OpenStack (`https://www.openstack.org`), OpenNebula (`http://opennebula.org`), Eucalyptus (`https://www.eucalyptus.com`) and CloudStack (`http://cloudstack.apache.org`), commonly offer elasticity through manual mechanisms (e.g., command line and graphical tool that allow users to control virtual machines). Complementary solutions such as Elastack [21], which provides automated monitoring and adaptation functions, can be integrated with OpenStack-like systems to provide dynamic infrastructure elasticity. However, it works only at the infrastructure level, *i.e.*, applications have to be made aware that nodes can be started or shut down at any time. In other words, it is up to the developers to ensure any kind of consistency or failure tolerance in the applications.

More recently, different research initiatives started to look at how elasticity can be leveraged by HPC applications. As an example, ElasticMPI proposes an elasticity framework for MPI applications through the stop-reconfigure-and-go approach [2]. However, this approach can negatively impact the performance of applications, in particular those that do not have long execution times. A second drawback of ElasticMPI is that it requires applications to be modified. Another approach, named Auto-elasticity [22], considers a pre-defined auto-elasticity by adjusting the number of VM instances accordingly to the application's input data (workload). In other words, as Auto-elasticity assumes that a program is modeled on deadline basis, the number of VMs is pre-defined in order to meet the deadlines.

Most of the existing solutions that provide cloud elasticity for high performance applications are commonly built around the master-slave programming model [2, 11, 23]. In case of iterative applications, which is the most common one, it means that at each new loop the master redistributes the tasks to slaves [2, 11]. However, in most cases the elasticity of the system is provided in a reactive way at the IaaS level, *i.e.*, without knowledge of on-the-fly information from the applications. Summing up, current approaches suffer from different issues such as (i) lack of mechanism to verify whether the application reached (or not) its peak load when achieving a load balancing threshold value [21, 23]; (ii) extra complexity at the application level, *i.e.*, the code needs to be instrumented and/or reorganized [2, 11]; (iii) static elasticity defined by pre-execution information [2, 14]; (iv) reconfiguration of the application's resources using a stop-and-relaunch approach [2]; and (v) assume that the communication latency between any two VMs is constant [24].

Considering the scope of MPI applications, Raveendran, Bicer and Agrawal [2] proposed one of the most advanced approaches to support the execution of such kind of applications. Nevertheless, as mentioned above, their solution needs application data in advance to feed the elasticity middleware and the insertion of elasticity code in the MPI application, besides the need to stop and relauching the whole application when elasticity takes place. Observing the initiatives described here, we are prooposing AutoElastic as a first step towards addressing the aforementioned issues (i), (ii), (iii), and (iv). In other words, our solution does not add any extra code or complexity to the existing HPC applications, allows dynamic (runtime) elasticity, and enables on-the-fly reconfiguration of resources without having to stop and relaunch the application.

3. AutoElastic Model

Traditionally, HPC applications are executed on clusters or even in grid architectures. In general, both have a fixed number of resources that must be maintained in terms of infrastructure configuration, scheduling (where tools such as PBS[2], OAR[3], OGS[4] are usually employed for resource reservation and job scheduling) and energy consumption. In addition, the tuning of the number of processes to execute a HPC application can be a hard procedure: (i) both short and large values will not explore the distributed system in an efficient way; (ii) a fixed value cannot fit irregular applications, where the workload varies along the execution and/or sometimes it is not predictable in advance. On the other hand, cloud elasticity abstracts the infrastructure configuration and technical details about resource scheduling from users, who pay for resources, and energy consequently, in accordance with the application's demands. However, the main gaps between the duet HPC and elasticity are application modeling and the overhead related to scaling out operations. Aiming at addressing these gaps, we propose AutoElastic – a cloud elasticity model that operates at the PaaS level of a cloud, acting as a middleware that enables the transformation of a non-elastic parallel application in an elastic one. Thus, AutoElastic was proposed as a solution to answer questions such as:

(i) *Is it possible to provide cloud elasticity to high performance computing applications in a transparent and non-intrusive way (i.e., without needing to modify applications)?*

(ii) *Which HPC applications can benefit from cloud elasticity and what are the gains of using cloud elasticity?*

(iii) *What are the minimal assumptions to transparently support cloud elasticity in HPC applications?*

[2] Project Website: http://www.arc.ox.ac.uk/content/pbs

[3] Project Website: https://oar.imag.fr/

[4] Previously known as Sun Grid Engine (SGE). Project Website: http://gridscheduler.sourceforge.net

AutoElastic provides transparent horizontal and reactive elasticity for parallel applications, *i.e.*, without requiring the intervention of the programmer (also named here as cloud user) to specify sets of rules, actions, or modify the application's code. Figure 3 (a) illustrates the traditional approaches of providing cloud elasticity to HPC applications, while (b) highlights AutoElastic's idea. The approach proposed by AutoElastic allows users to submit a traditional, non-elastic aware, application to the cloud, while the framework takes care of resource reorganization through automatic VM allocation and consolidation procedures. As AutoElastic works at the granularity of virtual machines, it has to be aware of the VM instantiation overhead to provide seamless elasticity, *i.e.*, in a non-prohibitive way for HPC applications.

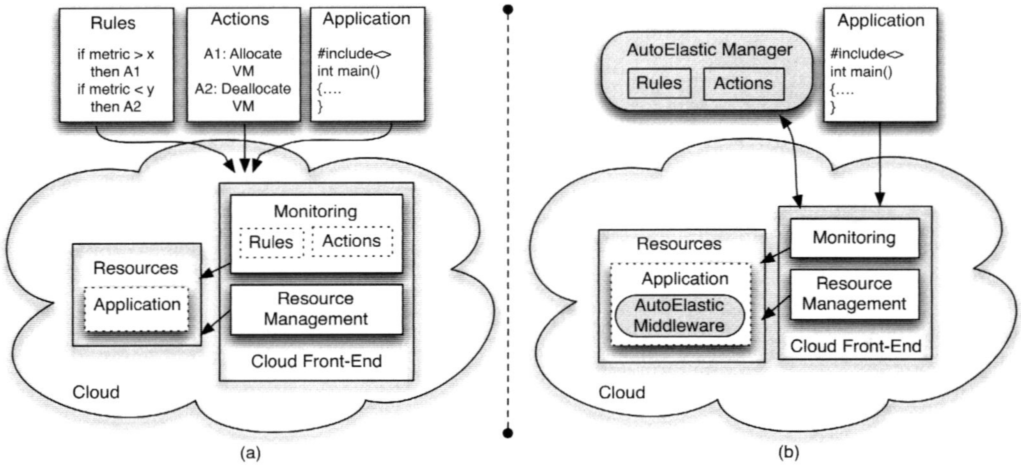

Figure 3. General ideas on using elasticity: (a) standard approach adopted by Amazon AWS and Windows Azure, in which the user must pre-configure a set of elasticity rules and actions; (b) AutoElastic idea, contemplating a manager that coordinates the elasticity actions and configurations on behalf of the user.

3.1. Architecture

AutoElastic is a middleware that operates as PaaS (Platform as a Service) that allows non-elastic parallel applications to take advantage of cloud elasticity without any change. To provide elasticity, it works with scaling in and out operations that consolidate or allocate virtual machine instances, respectively. Figure 4 depicts the AutoElastic architecture, presenting the framework components and the mapping of VMs. The framework includes a Manager, which can be either assigned to a virtual machine inside the cloud or to act as a stand-alone program outside the cloud. This is possible by taking advantage of cloud-supported APIs. As HPC applications are commonly CPU-bounded, we opted to create a process per VM and c working VMs per computing node, where c refers to the number of computational cores inside the node. This design decision has been previously investigated and validated as a way of exploring the efficiency of large computing nodes [25]. In addition, Figure 4 also presents the first ideas regarding the scope of HPC applications, presenting VMs that execute master and slave processes.

The AutoElastic Manager monitors the virtual machines, taking elasticity actions when considering them as pertinent for the current hardware and application behavior. The user can inform a file with an SLA (Service-Level Agreement) containing the minimum and the maximum number of allowed VMs to execute the application on the cloud. If no SLA is provided, the

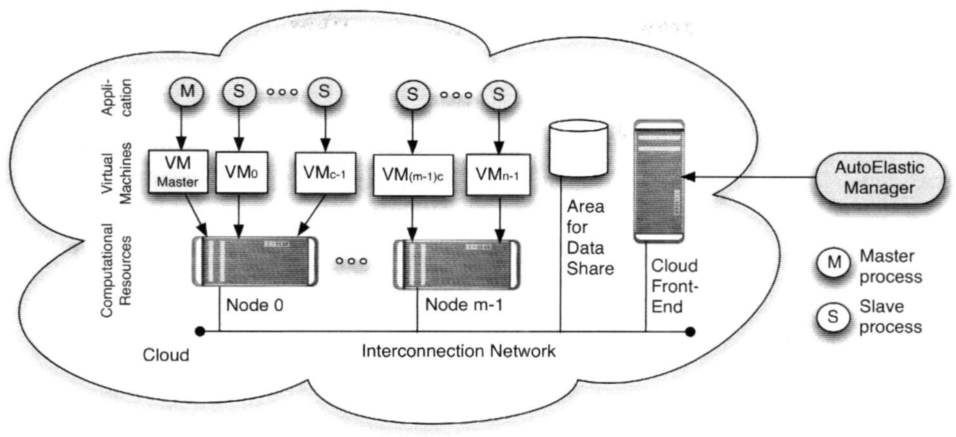

Figure 4. The AutoElastic architecture. While the number of nodes is m, the number of cores in a node is identified by c. The number of VMs running slave processes is n, which can be computed by $c \times m$.

default upper bound of virtual machines is two times the number of VMs used when launching the application. Instead of offering an application-sided elasticity, the use of a manager brings the benefits to resource reorganization in an asynchronous way at the application perspective, not penalizing it on VM (de)allocation actions. However, this non-blocking operation implies in the following question: *How can we notify the application about the resource reconfiguration?* We can achieve this goal through a framework that implements the concept of asynchronous elasticity.

> *Asynchronous elasticity* is a way of asynchronously notifying applications regarding changes on the underlying infrastructure, such as the number of computing instances. For instance, the application is notified as soon as a new computing VM instance (scale out) is available in the system without impairing its normal execution flow.

AutoElastic provides a framework that implements the concept of asynchronous elasticity. One of its key elements to provide asynchronous elasticity in a transparent fashion is a shared data area, which is used to provide interaction between the AutoElastic Manager and the VMs inside the cloud. Shared data areas are a common practice for sharing data between VM instances on cloud infrastructures [17, 18, 19]. They can be implemented by different means such as network file systems, message-oriented middlewares, and tuple spaces. Thus, AutoElastic uses the shared data area as a manner to combine HPC application and cloud elasticity, so providing actions as presented in Table 1.

Table 1. Actions provided through the shared data area.

Action	Direction	Description
Action 1	AutoElastic Manager → Master Process	There is a new resource with c virtual machines which can be accessed using given IP addresses.
Action 2	AutoElastic Manager → Master Process	Request for permission to consolidate a specific node, which encompasses given virtual machines.
Action 3	Master Process → AutoElastic Manager	Answer for Action 2 allowing the consolidation of the specified computing node.

The shared data area provides three types of notifications, as summarized in Table 1. Action 1 is an asynchronous notification sent by the AutoElastic Manager to the application announcing new ready to use computing resources. Figure 5 illustrates the functioning of the AutoElastic Manager when creating a new slave, so launching Action 1 afterwards. Action 2 is required for two reasons: (i) to avoid abruptly finishing a running process, which might lead to data losses; (ii) to ensure that the application will not be aborted due to a sudden interruption of a process. This second rationale is particularly important for MPI applications that execute over TCP/IP networks since they are usually aborted when a process abruptly disconnects. Finally, Action 3 is a decision taken by the master process that avoids inconsistent global state during the application's execution. In other words, once Action 2 has been received, the master process does not dispatch any task to the specific slaves which belong to the node that will be consolidated. The shared data area plays a key role in this process since it keeps all processes updated regarding any resource reconfiguration, allowing a safely adaptation to the new network topology.

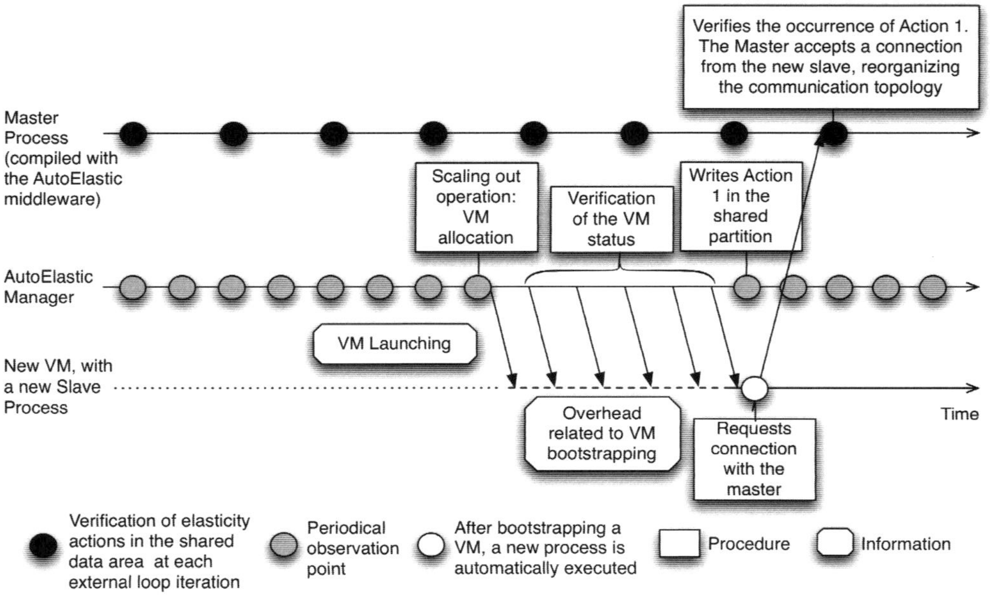

Figure 5. Functioning of the master, the new slave and the AutoElastic Manager to enable the Asynchronous Elasticity.

AutoElastic uses VM replication to provide cloud elasticity for HPC applications [26]. When scaling out the Manager launches new virtual machines using a pre-defined VM template. If the current nodes are working at full capacity, the Manager will first allocate a new computing node to launch the new VMs. The bootstrap of a VM is a time consuming procedure (e.g., boot time of the operating system) that finishes with the execution of a slave process. This slave automatically requests a connection to the master process, completing the asynchronous elasticity cycle. The master process will include the new slaves in the process group without any disruption or interruption on the application's execution. After that, the new slave processes will normally receive tasks from the master. The consolidation (scale in) takes place at a node granularity and not at the VM or process level. This design decision seeks to explore efficiency and energy saving, not using the power of a computing node partially. In fact, it has been claimed before that the number of VMs or processes inside a node is not the main factor for energy saving, but the fact that the node is turned on or off [27] .

Similarly to previous work [20, 28], AutoElastic performs resource monitoring periodically. Considering a monitoring interval, AutoElastic captures the CPU metric and computes a time series based on the lower and upper thresholds [29]. Particularly, thresholds are largely used in the state-of-the-art of cloud elasticity to drive resource reorganization on CPU-bound applications [1, 2, 4, 28]. AutoElastic uses the concept of moving average over a specific number of load observations to generate a single metric value; so elasticity actions are triggered on situations in which one of the thresholds violates this metric. To accomplish this, we are collecting CPU data using the function LP (Load Prediction), as presented in Equations 1 and 2. $MA(i,j)$ informs the CPU load of a virtual machine j at the observation number i. It performs a moving average considering the last x observations of load C, taking at start point the observation number i. Using this value, we compute an arithmetic average, so establishing an average load for the system by using the function $LP(i)$. In this function, n refers to the number of virtual machines in execution. Action 1 is triggered if LP is larger then the upper threshold, while Action 2 takes place when LP is shorter than the lower threshold. Finally, Equation 3 presents an empirical definition of the cost for execution an application with elasticity. The total number of observations is expressed by z, while $Active\_VMs(i)$ gives us the number of VMs in execution at observation i ($1 \leq i \leq z$). These numbers are important to compare elastic and non-elastic executions of HPC applications. Non-elastic executions have always the same number of VMs for all observations.

$$MA(i,j) = \frac{\sum_{k=i-x+1}^{i} Cjk}{x} \quad where \ \ i \geq x \tag{1}$$

$$LP(i) = \frac{\sum_{j=1}^{n} MA(i,j)}{n} \tag{2}$$

$$Cost = app\_time \ \times \ \sum_{i=1}^{z} Active\_VMs(i) \tag{3}$$

3.2. Model of Parallel Application

AutoElastic explores data parallelism on iterative message passing applications, which are modeled following the master-slave parallel programming model. This parallel programming model is extensively used in genetic algorithms, the Monte Carlo technique, geometric transformations for 2D and 3D images, asymmetric cryptography and SETI@home-like applications [2]. However, it is worth emphasizing that the framework allows the existing processes of the HPC application to know the identifier of the new instantiated processes, *i.e.*, enabling also a all-to-all communication topology. In other words, it means that AutoElastic supports also applications such as BSP and Divide-and-Conquer.

For developing the communication framework, we investigated the semantics and syntax of both MPI 1.0 and 2.0. While the former statically creates all processes at launching time, the latter supports dynamic process creation and on-the-fly reconfiguration of the connection topology. It means that MPI 2.0 is suitable for elastic environments. The AutoElastic parallel applications follow the MPMD (Multiple Program Multiple Data) principle, where master and slave processes have different executable codes. Each type of binary is mapped to a different VM template. The idea is to offer application decoupling for processes with different purposes, enabling flexibility and making the implementation of elasticity easier. Listing 1 presents a pseudocode of an AutoElastic-supported iterative application. The master code executes a series of tasks, capturing each one sequentially and parallelizing one-by-one to be processed on slave processes. This behavior can be observed in the external loop (line 2).

Currently, AutoElastic works with the following MPI 2.0-like communications directives: (i) publication of a connection port; (ii) looking for a server, taking as starting point a connection port; (ii) connection request; (iv) connection accept and; (v) disconnection request. Different from the approach in which a master launches processes using the so-called *spawn()* directive, AutoElastic acts in accordance with the second MPI 2.0 approach to support dynamic process creation: Sockets-based point-to-point communication. The launching of a new VM automatically entails the execution of a slave process, which requests a connection to the master automatically, as presented in Listing 2. Here, we emphasize that an AutoElastic-supported application does not need to necessarily rely on the MPI 2.0 API, but only follow the semantics of the communication directives.

Listing 1. Pseudo-language of the the master process

```
 1. size = initial_mapping(ports);
 2. for (j=0; j< total_tasks; j++){
 3.    publish_ports(ports, size);
 4.    for (i=0; i< size; i++){
 5.       connection_accept(slaves[i], ports[i]);
 6.    }
 7.    calculate_load(size, work[j], intervals);
 8.    for (i=0; i< size; i++){
 9.       task = create_task(work[j], intervals[i]);
10.        send_assync(slaves[i], task);
11.    }
12.    for (i=0; i< size; i++){
13.       recv_sync(slaves[i], results[i]);
14.    }
15.    store_results(slave[j], results);
16.    for (i=0; i< size; i++){
17.       disconnect(slaves[i]);
18.    }
19.    unpublish_ports(ports);
20. }
```

Listing 2. Pseudo-language of the of the slave process

```
master = lookup(master_address, naming);
port = create_port(IP_address, VM_id);
while (true){
   connection_request(master, port);
   recv_sync(master, task);
   result = compute(task);
   send_assync(master, result);
   disconnect(master);
}
```

Listing 3. Code to manage elasticity in the master process

```
int changes = 0;
if (Action == 1){
   changes += add_VMs();
}
else if (Action == 2){
   changes -= drop_VMs();
   allow_consolidation(); // enabling Action3
}
if (Action ==1 or Action == 2){
   reorganize_ports(ports);
}
size += changes;
```

The `initial_mapping` method (line 1 of Listing 1) is used by the master process to verify the execution configuration, which defines the initial setup of virtual machines, an identifier and the IP addresses of each process. Taking into account this information, the master knows the number of slaves and creates port names to receive connections from the slave processes. The communication happens asynchronously, where the master sends data to slaves in a non-blocking fashion but receives data from them synchronously. In fact, loop-based programs are convenient to implement cloud elasticity because it is easier to reconfigure the number of resources in the beginning of each iteration without changing the application semantics. Moreover, the job distribution loop is where the global consistent state of the system is kept.

The user must not insert any line about cloud elasticity in the code of the application. AutoElastic middleware manages the transformation of a non-elastic application into an elastic one at the PaaS level by one of the following strategies: (i) polymorphism can overload a method to provide elasticity for object-oriented implementations; (ii) a source-to-source translator can be used to insert code between the lines 1 and 2; (iii) a wrapper for the function in line 3 can be developed for procedural languages. Independent of the strategy, the code required for elasticity is simple, as shown in Listing 3. First, we need to verify if there is a new action from the AutoElastic Manager in the shared data area. If Action 1 has been activated, the master process reads the information regarding the new slaves and knows that it must expect new connections from them. In the case of Action 2, the master removes from its group the processes that belong to a specific node. After doing that, it triggers Action 3.

Although the design of AutoElastic takes into account master-slave applications, the iterative modeling and the use of MPI 2.0-like directives makes it easy to add and remove processes, as well as establishment completely new and arbitrary topologies. At the implementation level, it is possible to optimize connection and disconnection procedures if a particular slave process remains among the active ones in the process list. This improvement can benefit TCP-like connections that require a three way handshake protocol, which might be expensive for some applications.

4. Implementation

We developed an AutoElastic prototype for OpenNebula-based private clouds. The OpenNebula Java API, which was used for developing the AutoElastic Manager, provides the resources required to control both resource monitoring and scaling in and out activities. Moreover, the API is also used to launch parallel applications in the cloud. To run the processes, we created two VM templates, one for the master and another for the slaves. Following, we present some technical decisions in the prototype implementation:

- We used the WS-agreement XML standard[5] to define an SLA, which specify the minimum and maximum number of VMs for the tests;
- The shared data area was implemented through NFS, enabling all VMs inside the cloud infrastructure to access the files. The AutoElastic Manager, which can run outside of the cloud, uses the SSH protocol to access the shared data area on the front-end node;
- The load LP for the monitoring observation number i denoted $LP(i)$ is computed using the moving average of the slave VMs, considering an windows with 3 observations;
- The interval used for monitoring data was 30 seconds;
- Based on the related work (see Section 2), we defined 40% and 80% as the lower and upper thresholds, respectively.

5. Parallel Application and Evaluation Methodology

We developed a numeric integration application to evaluate the gains with and without asynchronous elasticity. The idea was to observe the benefits (e.g., gains in performance, such

[5] https://www.ogf.org/documents/GFD.192.pdf

as reduced execution time) of cloud elasticity for HPC applications. The application computes the numerical integration of a function *f(x)* in a closed interval $[a, b]$. In the implementation, we used the Composite Trapezoidal rule from a Newton-Cotes postulation [30]. The Newton-Cotes formula can be useful if the value of the integrand is given at equally spaced points. Considering the partition of the interval $[a, b]$ into s equally spaced subintervals, each one with length h ($[x_i; x_{i+1}]$, for $i = 0, 1, 2, ..., s-1$). Thus, $x_{i+1} - x_i = h = \frac{b-a}{s}$. The integral of *f(x)* is defined as the sum of the areas of the s trapezoids contained in the interval $[a, b]$, as presented in Equation 4. Equation 5 shows the development of the integral in accordance with the Newton-Cotes postulation.

$$\int_a^b f(x)\,dx \approx A_0 + A_1 + A_2 + A_3 + ... + A_{s-1} \tag{4}$$

where A_i = area of trapezoid i, with $i = 0, 1, 2, 3, ..., s-1$.

$$\int_a^b f(x)\,dx \approx \frac{h}{2}[f(x_0) + f(x_s) + 2.\sum_{i=1}^{s-1} f(x_i)] \tag{5}$$

The values of x_0 and x_s in Equation 5 are equal to a and b, respectively. In this context, s means the number of subintervals. Following this Equation, there are $s+1$ $f(x)$-like simple equations for obtaining the final result of the numerical integration. The master process must distribute these *s+1* equations among the slaves. Logically, some slaves can receive more work than others when *s+1* is not fully divisible by the number of slaves. Thus, the number of subintervals s define the computational load for each equation.

Aiming at analyzing the parallel application on different input loads, we considered four patterns: Constant, Ascending, Descending and Wave. Table 2 and Figure 6 show the equation of each pattern and the template used in the tests. The iterations in this figure mean the number of functions that are generated, resulting in the same number of numerical integrations. Additionally, the polynomial selected for the tests does not matter in this case because we are focusing on the load variations and not on the result of the numerical integration itself.

Table 2. Functions to express different load patterns. In *load(x)*, x is the iteration index at application runtime.

Load	Load Function	Parameters			
		v	w	t	z
Constant	$load(x) = \frac{w}{2}$	-	1000000	-	-
Ascending	$load(x) = x * t * z$	-	-	0.2	500
Descending	$load(x) = w - (x * t * z)$	-	1000000	0.2	500
Wave	$load(x) = v * z * sen(t * x) + v * z + w$	1	500	0.00125	500000

Figure 7 shows a graphical representation of each pattern. The x axis in the graph expresses the number of functions (one function per iteration) that are being tested, while the y axis informs the respective load. The load means the number of subintervals s between the limits a and b, which in this experiment are 1 and 10, respectively. The larger the number of intervals is, the greater the computational load for generating the numerical integration of the function. For the sake of simplicity, the same function is employed in the tests, but the number of subintervals for the integration varies. Considering the cloud infrastructure, OpenNebula is executed in a cluster with 10 nodes. Each node has two processors, which are exclusively dedicated to the cloud middleware. AutoElastic Manager runs outside the Cloud and uses the OpenNebula API to control and launch VMs. Our SLA was set up for a minimum of 2 nodes (4 VMs) and a maximum of 10 nodes (20 VMs).

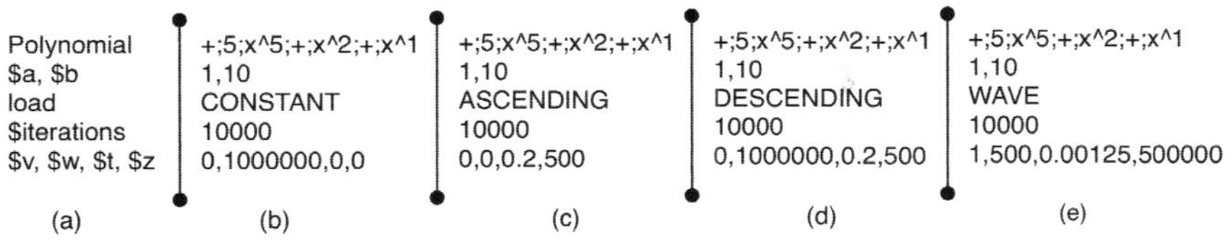

Figure 6. (a) Template of the input file for the tests; (b), (c), (d) and (e) are instances of the template when observing the load functions in Table 2.

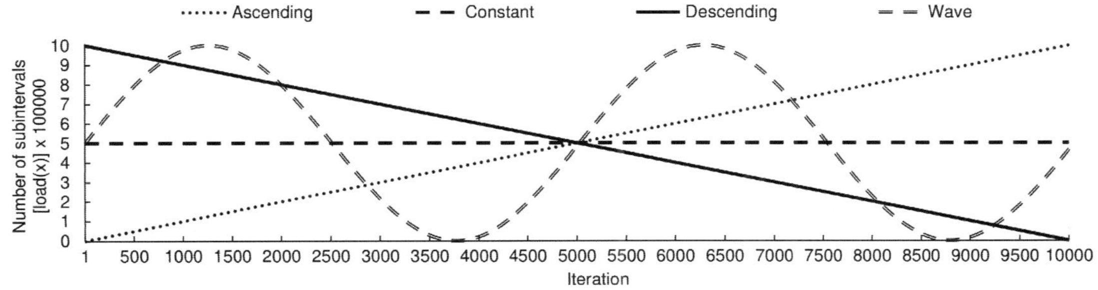

Figure 7. Graphical vision of the load patterns.

6. Evaluation and Discussing Results

We evaluated the numerical application using four load patterns in two scenarios: enabling and disabling cloud elasticity. At each execution, the initial configuration considers 2 nodes, the first executing 2 VMs (2 slave processes) and the second executing 3 VMs (2 slave processes and the master). We collected two metrics, the time (in seconds) to execute the application and the number of load observations performed by AutoElastic during the execution. At each observation i, we have the number of VMs execution on that moment, as well as the result for $LP(i)$. The results can be seen in Table 3. The last column shows the cost according to Equation 3.

Table 3. Results of the executions with and without elasticity support.

Elasticity	Load	Observations with			Total Observations	Time	Cost
		4 VMs	6 VMs	8 VMs			
Disabled	Ascending	84	0	0	84	2426	815136
	Constant	79	0	0	79	2370	748920
	Descending	84	0	0	84	2397	805392
	Wave	84	0	0	84	2444	821184
Enabled	Ascending	31	26	8	65	1978	680432
	Constant	79	0	0	79	2370	748920
	Descending	9	14	33	56	1775	681600
	Wave	9	29	22	60	1895	731470

As can be seen in the Table 3, when elasticity is enabled, the loads Ascending, Descending and Wave used different numbers of VMs during the application execution time. On the other hand, the load Constant used the same configuration in both scenarios with elasticity disabled and enabled. This behavior happened because the $LP(i)$ remained between the lower and upper thresholds, *i.e.*, no elasticity operations were necessary. On the other hand, the execution time and the amount of observations are lower in executions where resource reorganizations happened.

In the Ascending load with elasticity enabled, 47.7%, 40% and 12.3% of the observations returned 4, 6 and 8 VMs, respectively. The allocation of more VMs along the execution brings a better execution final when compared to the non-elastic execition. This happens because the load grows slowly and takes more time to the $LP(i)$ reach the upper threshold, thus the VM configuration stays with the initial configuration (4 VMs) for a long time. This behavior repeats itself when new resources are available. In the Descending case, 16.1%, 25% and 58.9% of the observations returned 4, 6 and 8 VMs respectively. Here the behavior is opposite to the Ascending load because, in this turn, the resources are allocated in the beginning of the execution and, as the load decreases slowly, it takes more time to reach the lower threshold. Finally, in the Wave load 15%, 48.3% and 36.7% of the observations returned 4, 6 and 8 VMs, respectively. In this case, as the load grows and decrease during the execution, it needs more resources in the beginning and after varies between the scenarios with 6 and 8 VMs.

Figures 8 and 9 illustrate the execution time of the application and the total cost obtained on each scenario. The elastic execution outperforms the non-elastic execution in the Ascending, Descending and Wave patterns, presenting performance gains of 18%, 26% and 22%, respectively. This behavior was also perceived when observing the cost, where AutoElastic with elasticity support resulted in costs approximately 14%, 11% and 10% lower than those with the non-elastic execution for the same mentioned load patterns. Considering that we are allocating more resources on-the-fly to avoid bottlenecks in the application's execution, elasticity helped to reduce the execution times, as can be seen in Table 3. Although using more resources, the gain in the time metric is enough to provide the lower values of cost in favor of the elastic execution. In other words, when compared with the non-elastic execution, AutoElastic uses more resources, which is compensated in terms of execution time.

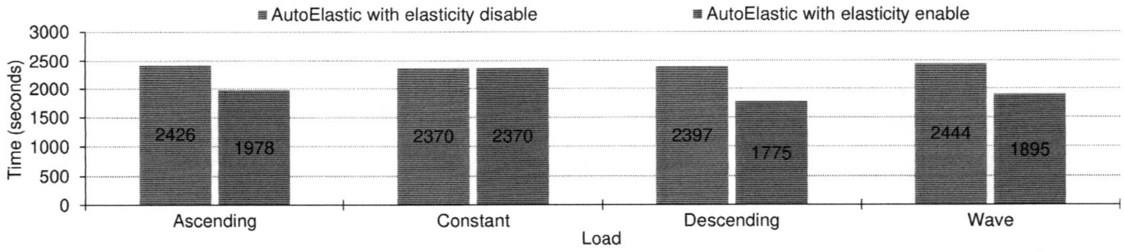

Figure 8. Time to execute the parallel application in the different scenarios and loads.

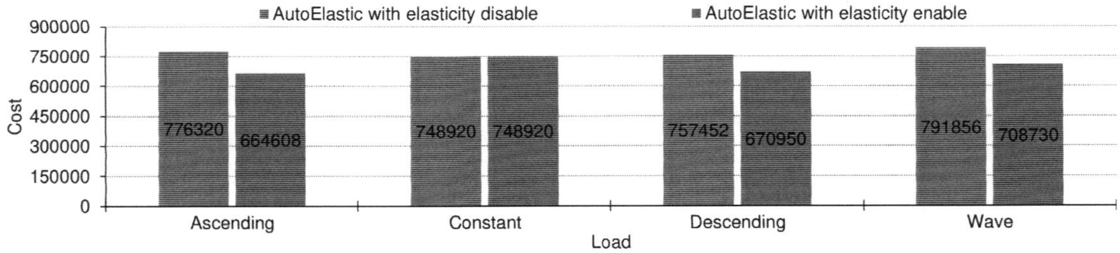

Figure 9. Cost obtained to execute the parallel application in the different scenarios and loads.

Figure 10 depicts a comparison regarding the history of resource allocation when combining load patterns and scenarios. We are not considering the Constant pattern because it does not cause elasticity actions. As expected, we allocate resources in specific moments in the Ascending pattern, while the Descending pattern shows a behavior of allocation in the beginning and a single deallocation in the end of the application. We leave as a future work a deeper analysis of the impact of variable thresholds. More precisely, Figure 10 (b) presents a situation in

Journal of Physics: Conference Series **649** (2015) 012006 doi:10.1088/1742-6596/649/1/012006

which resource management could be improved by increasing the value of the lower threshold. This strategy would imply in a better reactivity and resource usage, since the resources in the descending part will be deallocated sooner. Finally, parts (d), (e) and (f) of Figure 10 present executions in which the CPU demand is close to the theoretical rate available for the application, indicating moments of saturation and so, compromising the application performance.

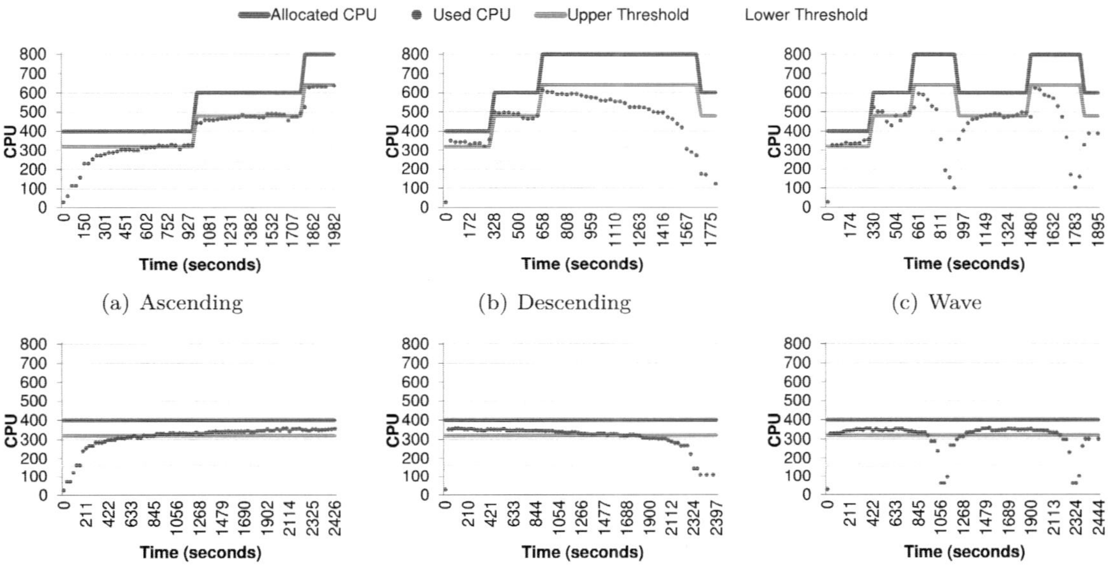

(a) Ascending (b) Descending (c) Wave

(d) Ascending (e) Descending (f) Wave

Figure 10. CPU behavior: (a), (b) and (c) with the elasticity enabled and (d), (e) and (f) with the elasticity disabled.

In the environment testbed, the procedure of allocating new resources comprises the transferring of two VMs to a new node in a 100 Mbps network and the initialization of the VMs afterwards. Each VM is based on a template with 700 MBytes in size. During the whole phase of allocating new VMs, the application executes normally with the current resources. The resource reorganization is performed only after completely delivering the new VMs. Table 4 presents the instants in time when new resources were allocated in the tests (see Figure 10). In this table, "VM allocation" represents the instant (including both number of the observation and application time in this instant) in which the $LP(i)$ function violates the threshold so triggering a new resource allocation. The term "VM delivering" represents the moment in which previously allocated resources were deallocated, *i.e.*, detached from the application. The average time between the start of a resource allocation and its delivering to the application is about 214 seconds.

As can be observed in Figure 10, when VMs are being delivered (Table 4), the accumulated CPU load automatically increases as more resources are available. This happens because AutoElastic only notifies the existence of new resources to the application when the the VMs are totally up avoiding pauses in the application execution since the application only needs to connect with the new processes. Figure 11 illustrates the amount of resources that are being delivered, as well as the amount of resources that are being allocated during the application execution. The specific instants in time can be see in Figure 10. Particularly, in blue we are emphasizing the current application resources and in green, we identify the resources that are

Table 4. Analyzing the the time interval between detecting the need to allocate and the delivery of 2 VMs at each elasticity action.

Load	Observation		Time (seconds)		Total Operation Time (seconds)
	VMs Allocation	VMs Delivering	VMs Allocation	VMs Delivering	
Ascending	25	31	752	958	206
Ascending	51	57	1562	1769	207
Descending	3	9	90	295	205
Descending	12	19	388	625	237
Wave	3	9	91	297	206
Wave	12	18	390	597	207
Wave	37	44	1211	1447	236

being allocated (see Table 4 for detail). This figure is important to observe that the resources are only delivered to the application after being completely up; meanwhile, the application executed normally without interruption, maintaining the current number of resources.

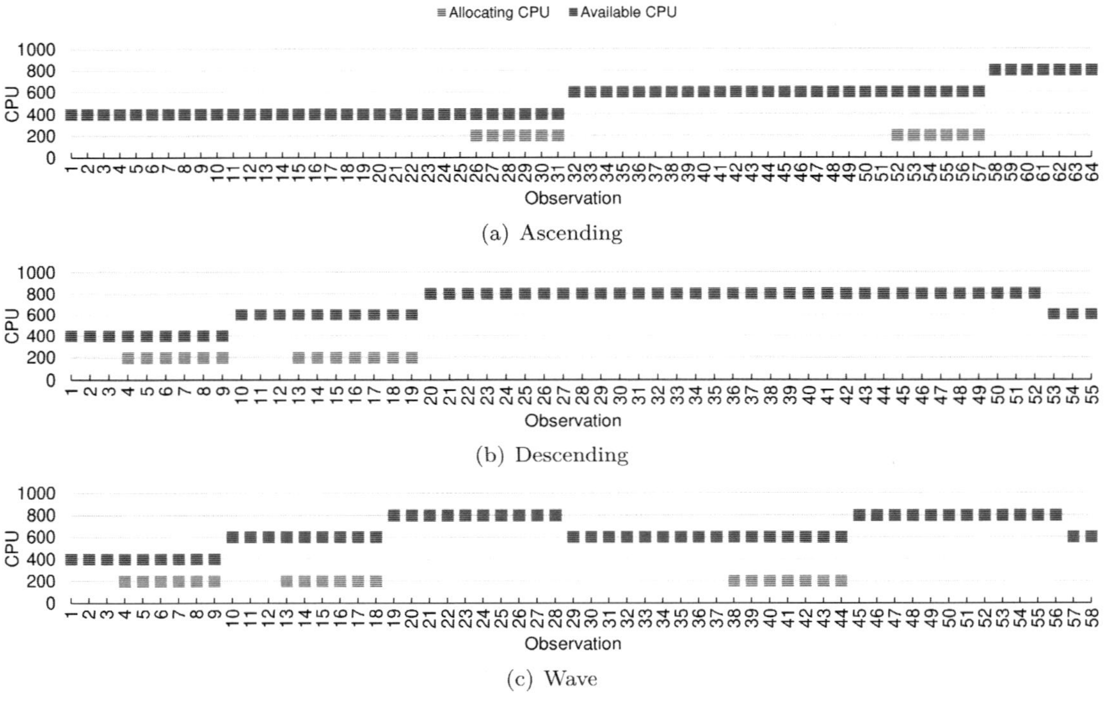

Figure 11. Resource allocations.

Figure 12 depicts the first resource allocation operation among those presented in Figure 11 (a). We can observe three things happening: (i) threshold violation, *i.e.*, the value of $LP(i)$ is greater than the upper threshold; (ii) instantiation of two VMs in a new node; (iii) delivery of the new allocated resources to the application.

7. Conclusion

This article addressed the cloud elasticity for iterative HPC applications through the proposition of the AutoElastic model. AutoElastic self-organizes the number of virtual machines without user intervention, bringing benefits both to the cloud administrator (better energy saving and

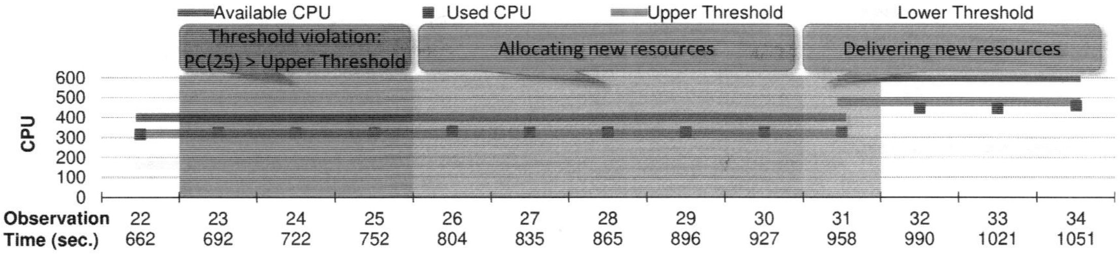

Figure 12. Detailed resource allocation process.

resource sharing among the users) and for the cloud users (who can take profit from a better performance and a quickly application deployment in the cloud). Section 3 presented three problem statements that were addressed as follows:

(i) AutoElastic acts at PaaS level, not requiring that the programmer write elasticity actions and rules in the application code to provide an elastic execution. It also offers asynchronous elasticity, which proved relevant to enable the use of HPC applications in the cloud computing environment.

(ii) The current version of AutoElastic works with master-slave iterative applications, not needing prior information about their behavior. AutoElastic provides a framework totally compatible with tightly-coupled applications, so models such as BSP and Divide-and-Conquer can be adapted in the future to take advantage of cloud elasticity. Concerning the performance gains with cloud elasticity, the evaluation showed that it is possible to reduce about 18% to 26% the execution time of a numerical integration application.

(iii) We are assuming that the user developed an iterative application, providing VM templates both for the master and the slave processes. Moreover, the user has an option to submit an SLA when launching the application. If not provided, AutoElastic takes as default twice the number of VMs at this time for the largest infrastructure.

Our approach for application model is justified by the fact that HPC programs can be developed with the Sockets-like MPI 2.0 programming style. This style allows process connection and disconnection easily, providing an effective use of available resources. AutoElastic offers a reactive and horizontal elasticity, going against the sentence claimed by Spinner et al. [31], who affirm that only vertical scaling is suitable for HPC scenarios due to inherent overhead related to the complementary approach. Thus, we modeled a framework to provide the novel concept of asynchronous elasticity, which turned out as a crucial feature to enable automatic resource reorganization without prohibitive costs. The aforesaid performance results are emphasized when analyzed together with the consumed energy, showing that the AutoElastic's elasticity does not present a forbidding cost.

As a future work, we intend to explore the self-organization of the thresholds in accordance with the application feedback. Finally, as explained earlier, we also plan to extend AutoElastic to contemplate other parallel programming models, including Divide-and-Conquer and BSP.

Acknowledgments

This work was partially supported by the following Brazilian Agencies: CNPq (Conselho Nacional de Desenvolvimento Científico e Tecnológico), CAPES (Coordenação de Aperfeiçoamento de Pessoal de Nível Superior) e FAPERGS (Fundação de Amparo à Pesquisa do Estado do Rio Grande do Sul).

References

[1] Lorido-Botran T, Miguel-Alonso J and Lozano J 2014 A review of auto-scaling techniques for elastic applications in cloud environments *Journal of Grid Computing* **12** pp 559–592
[2] Raveendran A, Bicer T and Agrawal G 2011 A framework for elastic execution of existing MPI programs *Proceedings of the 2011 IEEE Int. Symposium on Parallel and Distributed Processing Workshops and PhD Forum* IPDPSW '11 (Washington, DC, USA: IEEE Computer Society) pp 940–947
[3] Han R, Guo L, Ghanem M M and Guo Y 2012 Lightweight resource scaling for cloud applications *Cluster Computing and the Grid, IEEE International Symposium on* **0** 644–651
[4] Ward J S and Barker A 2014 Self managing monitoring for highly elastic large scale cloud deployments *Proceedings of the Sixth International Workshop on Data Intensive Distributed Computing* DIDC '14 (New York, NY, USA: ACM) pp 3–10
[5] Galante G and Bona L C E d 2012 A survey on cloud computing elasticity *Proceedings of the 2012 IEEE/ACM Fifth International Conference on Utility and Cloud Computing* UCC '12 (Washington, DC, USA: IEEE Computer Society) pp 263–270
[6] Jennings B and Stadler R 2014 Resource management in clouds: Survey and research challenges *Journal of Network and Systems Management* 1–53
[7] Frincu M E, Genaud S and Gossa J 2013 Comparing provisioning and scheduling strategies for workflows on clouds *Proceedings of the 2013 IEEE 27th International Symposium on Parallel and Distributed Processing Workshops and PhD Forum* IPDPSW '13 (Washington, DC, USA: IEEE Computer Society) pp 2101–2110
[8] Wilkinson B and Allen C 2005 *Parallel Programming: Techniques and Applications Using Networked Workstations and Parallel Computers* An Alan R. Apt book (Pearson/Prentice Hall)
[9] Roloff E, Birck F, Diener M, Carissimi A and Navaux P 2012 Evaluating high performance computing on the Windows Azure Platform *Cloud Computing (CLOUD), 2012 IEEE 5th International Conference on* pp 803–810
[10] Coutinho E, de Carvalho Sousa F, Rego P, Gomes D and de Souza J 2014 Elasticity in cloud computing: A survey *annals of telecommunications - annales des telecommunications* 1–21
[11] Rajan D, Canino A, Izaguirre J A and Thain D 2011 Converting a high performance application to an elastic cloud application *Proceedings of the 2011 IEEE Third International Conference on Cloud Computing Technology and Science* CLOUDCOM '11 (Washington, DC, USA: IEEE Computer Society) pp 383–390
[12] Knauth T and Fetzer C 2011 Scaling non-elastic applications using virtual machines *Cloud Computing (CLOUD), 2011 IEEE International Conference on* pp 468–475
[13] Kumar K, Feng J, Nimmagadda Y and Lu Y H 2011 Resource allocation for real-time tasks using cloud computing *Computer Communications and Networks (ICCCN), 2011 Proceedings of 20th International Conference on* pp 1–7
[14] Michon E, Gossa J and Genaud S 2012 Free elasticity and free CPU power for scientific workloads on IaaS clouds *Parallel and Distributed Systems (ICPADS), 2012 IEEE 18th International Conference on* pp 85–92
[15] Hendrickson B 2009 Computational science: Emerging opportunities and challenges *Journal of Physics: Conference Series* **180** 012013
[16] Tan L, Kothapalli S, Chen L, Hussaini O, Bissiri R and Chen Z 2014 A survey of power and energy efficient techniques for high performance numerical linear algebra operations *Parallel Computing* **40** 559–573
[17] Cai B, Xu F, Ye F and Zhou W 2012 Research and application of migrating legacy systems to the private cloud platform with Cloudstack *Automation and Logistics (ICAL), 2012 IEEE International Conference on* pp 400–404
[18] Milojicic D, Llorente I M and Montero R S 2011 OpenNebula: A cloud management tool *Internet Computing, IEEE* **15** 11–14
[19] Wen X, Gu G, Li Q, Gao Y and Zhang X 2012 Comparison of open-source cloud management platforms: OpenStack and OpenNebula *Fuzzy Systems and Knowledge Discovery (FSKD), 2012 9th International Conference on* pp 2457–2461
[20] Chiu D and Agrawal G 2010 Evaluating caching and storage options on the Amazon Web Services Cloud *Grid Computing (GRID), 2010 11th IEEE/ACM International Conference on* pp 17–24
[21] Beernaert L, Matos M, Vilaça R and Oliveira R 2012 Automatic elasticity in OpenStack *Proceedings of the Workshop on Secure and Dependable Middleware for Cloud Monitoring and Management* SDMCMM '12 (New York, NY, USA: ACM) pp 2:1–2:6
[22] Mao M, Li J and Humphrey M 2010 Cloud auto-scaling with deadline and budget constraints *Grid Computing (GRID), 2010 11th IEEE/ACM International Conference on* pp 41–48
[23] Martin P, Brown A, Powley W and Vazquez-Poletti J L 2011 Autonomic management of elastic services in the cloud *Proceedings of the 2011 IEEE Symposium on Computers and Communications* ISCC '11 (Washington, DC, USA: IEEE Computer Society) pp 135–140
[24] Zhang X, Shae Z Y, Zheng S and Jamjoom H 2012 Virtual machine migration in an over-committed cloud

Network Operations and Management Symposium (NOMS), 2012 IEEE pp 196–203

[25] Lee Y, Avizienis R, Bishara A, Xia R, Lockhart D, Batten C and Asanovic K 2011 Exploring the tradeoffs between programmability and efficiency in data-parallel accelerators *Computer Architecture (ISCA), 2011 38th Annual International Symposium on* pp 129–140

[26] Kouki Y, Oliveira F A d, Dupont S and Ledoux T 2014 A language support for cloud elasticity management *Cluster, Cloud and Grid Computing (CCGrid), 2014 14th IEEE/ACM International Symposium on* pp 206–215

[27] Baliga J, Ayre R, Hinton K and Tucker R 2011 Green cloud computing: Balancing energy in processing, storage, and transport *Proceedings of the IEEE* **99** 149–167

[28] Imai S, Chestna T and Varela C A 2012 Elastic scalable cloud computing using application-level migration *Proceedings of the 2012 IEEE/ACM Fifth International Conference on Utility and Cloud Computing* UCC '12 (Washington, DC, USA: IEEE Computer Society) pp 91–98

[29] Jamshidi P, Ahmad A and Pahl C 2014 Autonomic resource provisioning for cloud-based software *Proceedings of the 9th International Symposium on Software Engineering for Adaptive and Self-Managing Systems* SEAMS 2014 (New York, NY, USA: ACM) pp 95–104

[30] Comanescu M 2012 Implementation of time-varying observers used in direct field orientation of motor drives by trapezoidal integration *Power Electronics, Machines and Drives (PEMD 2012), 6th IET International Conference on* pp 1–6

[31] Spinner S, Kounev S, Zhu X, Lu L, Uysal M, Holler A and Griffith R 2014 Runtime vertical scaling of virtualized applications via online model estimation *Proceedings of the 2014 IEEE 8th International Conference on Self-Adaptive and Self-Organizing Systems (SASO)*

ComprehensiveBench: a Benchmark for the Extensive Evaluation of Global Scheduling Algorithms

Laércio L Pilla[1], Tiago C Bozzetti[2], Márcio Castro[1], Philippe O A Navaux[2] and Jean-François Méhaut[3]

[1]Department of Informatics and Statistics, Technology Center, Federal University of Santa Catarina, Florianópolis, BR
[2]Informatics Institute, Federal University of Rio Grande do Sul, Porto Alegre, BR
[3]University of Grenoble Alpes, LIG, CEA-INRIA, Grenoble, FR

E-mail: laercio.pilla@ufsc.br, tcbozzetti@inf.ufrgs.br, marcio.castro@ufsc.br, navaux@inf.ufrgs.br, jean-francois.mehaut@imag.fr

Abstract. Parallel applications that present tasks with imbalanced loads or complex communication behavior usually do not exploit the underlying resources of parallel platforms to their full potential. In order to mitigate this issue, global scheduling algorithms are employed. As finding the optimal task distribution is an NP-Hard problem, identifying the most suitable algorithm for a specific scenario and comparing algorithms are not trivial tasks. In this context, this paper presents ComprehensiveBench, a benchmark for global scheduling algorithms that enables the variation of a vast range of parameters that affect performance. ComprehensiveBench can be used to assist in the development and evaluation of new scheduling algorithms, to help choose a specific algorithm for an arbitrary application, to emulate other applications, and to enable statistical tests. We illustrate its use in this paper with an evaluation of Charm++ periodic load balancers that stresses their characteristics.

1. Introduction

Science has advanced in the last decades partially by virtue of numerical simulations. These scientific applications are developed using parallel programming languages and interfaces in order to benefit from the computing and memory resources available in High Performance Computing (HPC) platforms. These applications are decomposed into parallel tasks (such as threads or processes) that are distributed over the available resources. Due to the nature of the simulations, tasks can possess different computational loads, complex communication graphs, or both. These behaviors result in load imbalance and communication overhead that affect the applications' performance and scalability. In these scenarios, global scheduling algorithms such as load balancers are employed to distribute tasks in a manner to mitigate performance issues [1].

As the problem of finding an optimal task distribution is known to be NP-Hard [2], several load balancing, work stealing, and task mapping algorithms have been proposed for different scenarios [3, 4, 5, 6, 7, 8, 9, 10, 11, 12]. These algorithms show different complexities, consider different information, and optimize or improve different objectives. Due to all this diversity, the comparison between different global scheduling algorithms is not a trivial task. This affects several questions related to the problem, including the following and their combination:

- *Is one scheduling algorithm better than another? Do we have statistically significant results to verify that?*
- *Should my new algorithm consider this information? Does it improve its task distribution?*
- *What is the best algorithm for my application? What is the best algorithm for my parallel platform?*
- *Should I migrate my application to an environment that supports task rescheduling?*

Current benchmarks used for the evaluation of scheduling algorithms lack a standard, usually limiting or ignoring some characteristics of scientific applications that influence algorithms' performance. Two examples of characteristics left out are the amount of data that has to be migrated with a task and the dynamic behavior of a task's load.

In this context, we propose a novel benchmark for the evaluation of global scheduling algorithms on shared and distributed memory platforms named ComprehensiveBench[1]. It considers the main characteristics of parallel applications that affect global scheduling algorithms' performance, and is more expressive in the way these characteristics are specified, enabling the simulation of dynamic behavior. To illustrate its capability of stressing the main characteristics of global scheduling algorithms, we perform experiments with ComprehensiveBench in different scenarios and load balancing algorithms. Our results confirm that ComprehensiveBench can be used to support the development and evaluation of scheduling algorithms.

The remaining sections of this paper are organized as follows: related work is discussed in Section 2. ComprehensiveBench is characterized in Section 3. Its implementation in Charm++ is presented in the same section. The experimental setup used to exemplify ComprehensiveBench's use and the global scheduling performance results are shown in Section 4. Finally, concluding remarks and future work are discussed in Section 5.

2. Related Work

Benchmarks have been extensively used to evaluate the performance of parallel platforms, parallelization tools, parallel libraries, and scheduling algorithms [3, 4, 5, 6, 7, 8, 9, 10, 11, 12]. Examples include the NAS Parallel Benchmarks (NPB) [13] and their multi-zone versions (NPB-MZ) [14], the Princeton Application Repository for Shared-Memory Computers (PARSEC) [15], the HPC Challenge benchmark suite (HPCC) [16], and the Rodinia benchmark suite [17]. Nevertheless, these benchmarks have limitations that impair the evaluation of global scheduling algorithms. For instance, NPB is limited to preset input sizes and is mostly useful for the evaluation of thread and process mapping algorithms that focus on improving communication performance, PARSEC and Rodinia can only be used in shared memory environments, and HPCC does not show load imbalance, focusing only in stressing parts of a parallel platform. The Mantevo Project [18] shows more flexibility with its miniapps, as MiniFE can be tuned to present load and communication imbalance, and Prolego can be defined using different code fragments. Nevertheless, dynamic features do not seem to be implemented.

Different approaches have been used to evaluate and compare global scheduling algorithms. When the main focus of the work is to improve one application's performance, only said application is used for comparison. This is the case for applications such as NAMD [4], BRAMS [5], Betweenness Centrality [7], and Ondes3D [19]. Another approach involves the generation of artificial application data for the offline comparison of algorithms, as done by

[1] Available online in the HieSchella project page: `http://forge.imag.fr/projects/hieschella/`

Catalyurek et al. [3]. A more toilsome way to compare algorithms requires the implementation of specific methods as benchmarks, as done by Lifflander, Krishnamoorthy and Kale [20]. Lastly, a more broad approach benefits from existing global scheduling benchmarks and mini-apps – such as *Adaptive Mesh Refinement* (AMR), *kNeighbor*, *lb_test*, *LeanMD*, and *stencil4D* – for experiments [6, 8, 9, 10]. Still, these benchmarks limit the expression of load imbalance, dynamic load behavior, communication behavior, initial mapping, size of tasks, or a combination of these. The use of benchmarks with limited ranges of parameters and behaviors jeopardizes the attainment of statistically significant results [21]. We explain how ComprehensiveBench circumvents these issues in the next section.

3. ComprehensiveBench

We describe in this section the main characteristics of our novel global scheduling benchmark. First, we discuss its main desired characteristics and features. This is followed by a description of all ComprehensiveBench's parameters with examples of use. Lastly, we detail how the benchmark is implemented.

3.1. Features

ComprehensiveBench displays a series of different characteristics to enable the comprehensive evaluation of global scheduling algorithms and emulation of applications. They include the expression of the following attributes:

- *Irregular loads*: tasks can have different loads.
- *Dynamic loads*: a task's load can change during execution.
- *Mixed loads*: tasks can execute integer or floating point operations.
- *Irregular communication*: the number of messages sent by a task and their sizes can vary among tasks.
- *Dynamic communication*: the number of messages sent by a task and their sizes can change during execution.
- *Diverse communication graphs*: tasks can be set to communicate with others following different patterns.
- *Irregular task sizes*: tasks can have different memory footprints.
- *Diverse initial mappings*: tasks can be initially mapped to cores following different patterns.

3.2. Parameters

ComprehensiveBench's parameters can be divided into two categories: simple and complex. *Simple parameters* are defined using natural numbers or keywords, while *complex parameters* are defined using expressions involving arithmetic operations, variables, numbers, and conditional (ternary) operators. We discuss each category in the following sections.

3.2.1. Simple Parameters ComprehensiveBench's static characteristics are all set using natural numbers or special keywords. Table 1 lists these benchmark characteristics. The number of tasks to be executed by ComprehensiveBench (`n`) directly affects the execution time of global scheduling algorithms, as a larger number of tasks leads to more time spent on mapping decisions. The number of cores to be considered by the benchmark (`p`) has a similar effect, as the larger the decision space is, the longer a scheduling algorithm may take to decide where to map a task.

The number of benchmark iterations (`r`) affects for how long it may execute. The longer ComprehensiveBench executes, the more time scheduling decisions have to affect

Journal of Physics: Conference Series **649** (2015) 012007 doi:10.1088/1742-6596/649/1/012007

Table 1. ComprehensiveBench's simple parameters.

Parameters	Meaning
`n`	Number of tasks.
`p`	Number of processing units (cores).
`r`	Number of iterations.
`lbfreq`	Rescheduling (load balancing) frequency.
`int_op`	Data type to be processed.
`gcomm`	Communication graph.

performance. The rescheduling frequency (`lbfreq`) has an effect related to it. If the rescheduling periodicity is small (few iterations), then performance issues like load imbalance and a high communication overhead can be fixed quickly. Nevertheless, if the algorithm takes too long computing a new task distribution each time, then its overhead may overcome its benefits. Meanwhile, if the rescheduling periodicity is large (many iterations), then it may take too long to correct a poor task distribution.

The `int_op` parameter is used to set the data type to be processed by tasks. If set to true, tasks work with integer data, otherwise tasks execute single precision floating-point operations. This parameter can be used to stress processors and influence the decisions of temperature- or energy-aware global scheduling algorithms.

The last simple parameter is the communication graph (`gcomm`). While the previous parameters are set using natural numbers mostly, `gcomm` is based in three predefined graphs: a ring, a 2D mesh, and a 3D mesh. The communication graph defines which tasks communicate with each other, and is independent of the machine topology. While the communication graph defines which tasks communicate with each other, it does not define how many messages will be exchanged nor their sizes. These are left to complex parameters discussed next.

3.2.2. Complex Parameters ComprehensiveBench's complex parameters are expressed using simple parameters and two different variables with values set at run time. These variables are presented in Table 2. By using a task's identifier (`t`) to influence its load and communication, ComprehensiveBench enables irregularity among tasks. Meanwhile, the use of the current iteration (`i`) makes it possible to vary loads and communication throughout the benchmark execution, bringing dynamicity. In this sense, the complex parameters work as functions that have to be computed for each task when first creating them or by each task at each iteration.

Table 2. ComprehensiveBench's run time variables.

Variable	Meaning	Range
`t`	Task identifier.	$0 \leq \mathtt{t} < \mathtt{n}$
`i`	Current iteration.	$0 < \mathtt{i} \leq \mathtt{r}$

Table 3 lists the complex parameters used in the benchmark, including their unities and set of variables and simple parameters used in their expressions. All functions result in integer values. For instance, the initial mapping of tasks to cores (`initmap`) can be defined using the task identifier, the total number of tasks being executed, and the total number of cores. Equation 1 shows the definition of a Round-Robin task distribution, while Equation 2 shows a compact task distribution, where a range of tasks are mapped to the same core, and Equation 3 shows a mapping where all tasks start executing in the same core. In all cases, `initmap` must be in the range $0 \leq \texttt{initmap} < \texttt{p}$. By changing the initial task mapping, ComprehensiveBench enables the creation of different scenarios with poor work and communication distributions to be improved by global schedulers.

$$\texttt{initmap} = \texttt{t} \mod \texttt{p} \tag{1}$$

$$\texttt{initmap} = (\texttt{t} * \texttt{p})/\texttt{n} \tag{2}$$

$$\texttt{initmap} = 5 \tag{3}$$

Table 3. ComprehensiveBench's complex parameters.

Parameters	Meaning	Variables and Parameters	Unit
`initmap`	Initial mapping.	`t, n, p`	Core id
`tasksize`	Task memory footprint.	`t, n`	Bytes
`load`	Task execution time.	`t, i, n`	ms
`msgnum`	Number of messages.	`t, i, n`	—
`msgsize`	Data volume per message.	`t, i, n`	Bytes

Each task occupies some space in memory associated with its code, internal variables and data structures. Whenever a task is assigned to a core different from its current one, its data has to be migrated. In this sense, a task's memory footprint has direct influence over its migration overhead and the scheduling algorithm's execution time. This data volume is expressed in bytes with the `tasksize` parameter. Its value is limited by the maximum integer possible in the platform (this is the same for `load`, `msgnum` and `msgsize`). All tasks can either occupy the same amount of memory, as shown in Equation 4, or have different memory footprints. The latter is exemplified in Equation 5, where tasks with higher identifiers have more data. A global scheduling algorithm that is aware of tasks' data volumes may opt to not migrate a task based in its estimated migration overhead.

$$\texttt{tasksize} = 1000 \tag{4}$$

$$\texttt{tasksize} = 100 * \texttt{t} * \texttt{t} \tag{5}$$

The execution time of a task is set to each of the benchmark's iterations by the `load` parameter. Its expression is evaluated at the beginning of each iteration in order to enable dynamicity. For instance, Equation 6 shows a situation where the load of all tasks increases by 10 ms at each iteration. Tasks can also be split into groups with different loads, as illustrated by the use of the conditional operator in Equation 7. A combination of dynamic and irregular loads is presented in Equation 8. Using this powerful expression, ComprehensiveBench is able to create varied load imbalance scenarios to stress global scheduling algorithms and to simulate the behavior of real applications.

$$\texttt{load} = 10 * \texttt{i} \tag{6}$$

$$\texttt{load} = \texttt{t} < \texttt{n}/2 \ ?\ 10 : 50 \tag{7}$$

$$\texttt{load} = 5 * \texttt{i} * (\texttt{t} + 1) \tag{8}$$

The last complex parameters are the number of messages to be sent to another task (`msgnum`) and the size of the messages in bytes (`msgsize`). By setting the communication behavior of tasks independently from one another, ComprehensiveBench provides the opportunity to test the communication resources of parallel platforms (i.e., memory and network) and the effectiveness of global scheduling algorithms in reducing the overall communication time of applications. Equation 9 illustrates a scenario where the number of messages increases with the execution of the benchmark, while Equation 10 simulates a scenario where tasks exchange more data at every five iterations.

$$\texttt{msgnum} = 1 + \texttt{i}/5 \tag{9}$$

$$\texttt{msgsize} = \texttt{i} \mod 5 < 1 \;?\; 10000 : 100 \tag{10}$$

3.3. Implementation Details

ComprehensiveBench is implemented using Charm++ [22, 23]. Charm++ is a parallel programming language and runtime system (RTS) based on C++ with the goal of improving programmer productivity. It abstracts architectural characteristics from the developer and provides portability over multiprocessor and multicomputer platforms. Charm++ applications are composed of specially designed C++ objects named *chares* that play the role of tasks, being responsible for their own work and data. Chares communicate using asynchronous method invocations following a message-driven execution.

Charm++ includes a load balancing framework [6] that provides an interface for developing load balancing plugins to Charm++ applications. A load balancer is provided with information regarding the last iterations of the application and its current mapping, and the runtime system expects in return a new task distribution to migrate tasks. Several global scheduling algorithms are distributed with this framework, which helps experimenting with our benchmark.

In order to enable the use of complex expressions to define the tasks' loads and other characteristics, all of ComprehensiveBench's parameters are set in an input file that is preprocessed right before the benchmark is compiled.

At the start of ComprehensiveBench's execution, all tasks are created and distributed among cores following the initial mapping expression. Their communication graph is also set during initialization. Since several tasks execute in the same core during an iteration, their loads cannot be simulated by just putting them to sleep. To overcome this issue, a task in the first core of the platform is responsible for estimating how many internal iterations must be done for a task to compute for a millisecond. This value is then used by all tasks to guide their loads at each iteration. This standardization of the workload also enables using ComprehensiveBench in environments with different processors or processors running at different clock frequencies.

4. Experimental Evaluation

In this section, we present experiments performed with ComprehensiveBench that illustrate its capability of stressing the characteristics of global scheduling algorithms. First, we detail the experimental setup with the parallel platform and Charm++ load balancing algorithms used. This is followed by the description and explanation of the results of four different experimental configurations.

4.1. Experimental Environment

Table 4 describes the parallel platform and system software used in the experiments. In total, 128 cores were used in the experiments. It is important to emphasize that this machine has a nonuniform memory access (NUMA) architecture with NUMA factors (ratio between remote latency and local latency) between 6.4 and 9.6. This results in slow communication between different NUMA nodes.

In the interest of reducing variability in the results, threads were pinned to cores in all experiments using `numactl`.

Table 4. Experimental platform.

Processors	16× Intel Xeon E5-4640
Cores	16 × 8 @ 2.4 GHz
Memory	16 × 32 GB DDR3 @ 1600 MHz
Linux kernel version	3.0.1010.46
G++ version	4.3.4
CHARM++ version	6.6.0
CHARM++ build	Multicore-linux64

4.2. Global Scheduling Algorithms

Five different load balancing algorithms implemented in CHARM++ were used in the experiments. Their main characteristics are discussed below.

- ***GreedyLB***: A centralized greedy algorithm that only uses task loads for its decisions. It is employed to quickly mitigate load imbalance. It sorts tasks in decreasing load order and iteratively maps the unassigned task with the highest load to the least loaded core. The initial task distribution is not considered in this process, which leads to a large number of task migrations.
- ***GreedyCommLB***: Similar to *GreedyLB* with the addition of communication information in its decisions. The algorithm computes a task's communication load based on the data volume that it sends to tasks mapped in other cores. After sorting tasks in decreasing load order, *GreedyCommLB* selects the unassigned task with the highest load and maps it to the core with the smallest total load, which includes both processing and communication loads. The algorithm has the tendency to map tasks with intense communication to the same core.
- ***RefineLB***: A less aggressive algorithm than the greedy strategies. The algorithm considers the current task mapping in its decisions, iteratively refining it to a state where no core is overloaded or no task migration improves it. *RefineLB* uses a threshold to define if a core is overloaded. At each iteration, it verifies all possible task migrations from the most overloaded core to all underloaded cores, migrating the task that brings its new core the closest to the threshold. As *GreedyLB*, it does not consider communication in its decisions.
- ***RefineCommLB***: Similar to *RefineLB*, but considers the communication behavior of tasks in its decision of the most suitable destination core for a task.
- ***RandCentLB***: Random algorithm that does not take into consideration any information about the application or the current state of the platform. It maps tasks to cores randomly using a uniform distribution.

Besides these five algorithms, all experiments were also executed without any load balancer to provide a baseline for the results.

4.3. COMPREHENSIVEBENCH *Parameters*

Four different synthetic setups were used with COMPREHENSIVEBENCH: the first one illustrates the impact of tasks' sizes on performance; the second one presents scenarios with static load imbalance; the third one shows how different algorithms handle dynamic load imbalance; and the forth one stresses how load balancers handle communication-intense applications on a NUMA platform. The parameters used in all three setups are listed in Table 5. All results presented in the next section represent the average times measured in 32 runs and present a statistical confidence of 95% by Student's t-distribution and a 2% relative error.

Journal of Physics: Conference Series **649** (2015) 012007 doi:10.1088/1742-6596/649/1/012007

Table 5. ComprehensiveBench parameters for the different scenarios tested.

Parameter	1st scenario	2nd scenario	3rd scenario	4th scenario
n	2000	2000	2000	2000
r	50	20	20	20
lbfreq	5	5	5	5
int_op	false	false	false	false
gcomm	Ring	Ring	Ring	Ring
initmap	(t * p)/n	t mod p	t mod p	(t * p)/n
tasksize	{10 KB, 100 KB, 1 MB}	100	100	100
load	10	$10 + \{0, 1, 2, 4\} *$ (t mod p)	$10 + ((i/10) * \{1, 2, 4\} *$ (t mod p)	10
msgnum	1	1	1	$\{1, 5, 10\}$
msgsize	100	100	100	10000

4.4. Results

The results of the experiments with each of the four scenarios are presented in the following sections.

4.4.1. Impact of Tasks' Sizes This first scenario simulates a balanced application whose tasks have nontrivial memory footprints, resulting in tasks with high migration costs. Tasks are set with small loads and little communication, and no benefit is expected to be achieved with load balancing. In this scenario, ComprehensiveBench helps evaluate the capacity of the different load balancing algorithms in estimating and avoiding migration costs.

The total execution times achieved by the different load balancing algorithms for tasks with sizes of 10 KB, 100 KB, and 1 MB are presented in Figure 1. The baseline represents the execution of ComprehensiveBench without any load balancer and its total execution time stays the same for all task sizes as no tasks are migrated. The increase in migration overhead with the increase of task size can be seen for load balancers *GreedyLB*, *GreedyCommLB*, and *RandCentLB*. These load balancers do not take into consideration the current task mapping and result in several task migrations at each load balancing call. The total migration overhead for these load balancers is approximately 4.6, 7, and 18 seconds for the increasing task sizes. This is a result of migrating 99% of the tasks at each load balancing call. *RandCentLB* shows a bigger increase in total execution time because its random migrations also result in load imbalance.

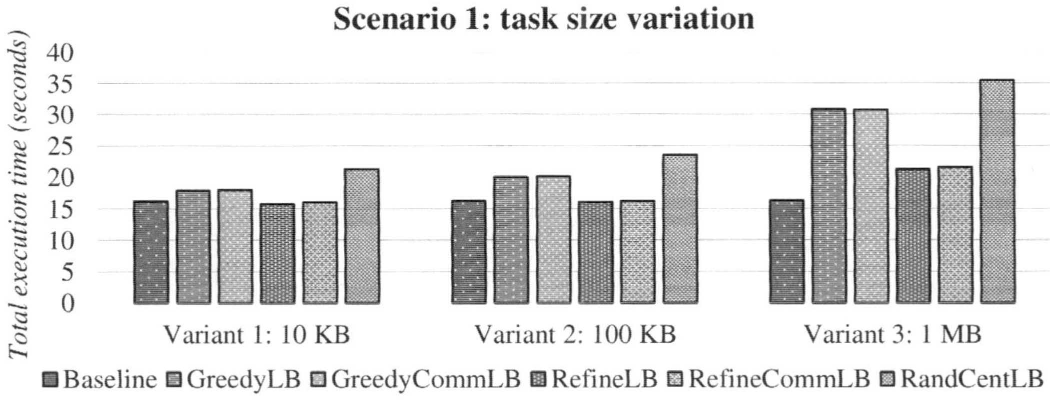

Figure 1. Load balancers' performance with varying task sizes.

Meanwhile, *RefineLB* and *RefineCommLB* achieved the same total execution time as the baseline for the two variants with smaller tasks, increasing execution time only when migrating 1 MB tasks. As these algorithms try to improve the current task mapping, they are able to notice that the load is mostly balanced, suffering only with some communication overhead and OS jitter, and thus, resulting in almost no migrations. This indicates that refinement-based algorithms can be much more suitable for application with large memory footprints or parallel platforms with high migration costs.

4.4.2. Influence of Static Load Imbalance The second scenario simulates an imbalanced application whose tasks have static loads. The load imbalance comes from having tasks in 128 load groups, where each group has tasks with the same load mapped to the same core. Tasks' sizes and communication are kept at a minimum to avoid influencing the final performance. The use of ComprehensiveBench helps to identify the possible performance gains with different algorithms for applications with irregular but static behavior, and how quickly these algorithms are able to mitigate load imbalance.

Figure 2 presents the total execution time with global scheduling algorithms and different levels of load imbalance. In Variant 1, all tasks and cores have the same load. The total execution time without load balancing and with all algorithms except *RandCentLB* is approximately 6.8 seconds, including iterations, communication, runtime overhead and initialization. Even though greedy algorithms migrate many tasks at each load balancing call, this migration overhead only increases total execution time by 0.6 seconds.

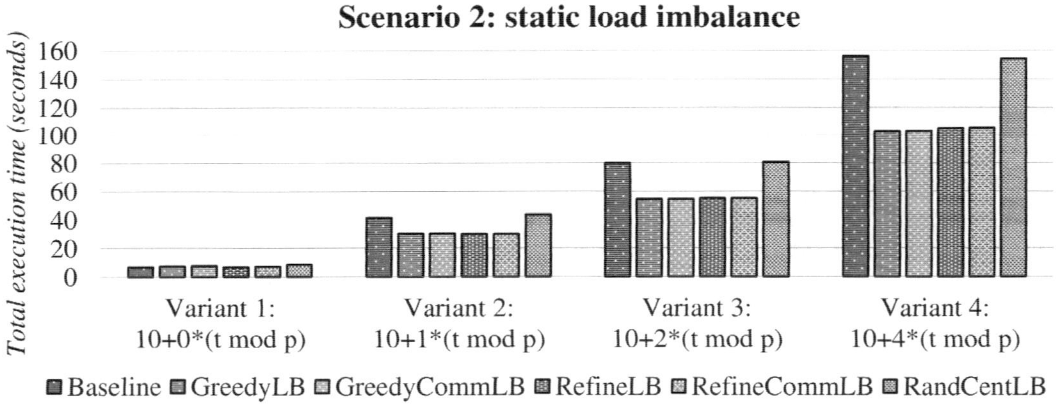

Figure 2. Load balancers' performance with static load imbalance.

As the load imbalance increases with the variant numbers, the total execution time of ComprehensiveBench increases accordingly. *GreedyLB*, *GreedyCommLB*, *RefineLB*, and *RefineCommLB* are able to improve performance in all variants, achieving almost optimal work distributions in all cases. The most visible difference in performance between greedy and refinement-based algorithms happens in Variant 4. Nevertheless, this difference is only 3% and is inside the error margin of the results.

Finally, it is interesting to see that even *RandCentLB* is able to improve performance (but still inside the error margin) as the initial load imbalance increases. This happens because the chances of having a mapping better than the original one increase as the difference between the most and least loaded cores increases.

4.4.3. Influence of Dynamic Load Imbalance The third scenario simulates an application that starts balanced but becomes imbalanced by the middle of its execution. This load imbalance is a result of dynamic changes in the tasks' loads, which is a common behavior in some scientific applications [5, 19]. In addition, load imbalance may happen even after several load balancing calls. In these situations, online global scheduling algorithms become necessary to achieve performance and scalability. The possibility of expressing dynamic behaviors is one of the main benefits of using ComprehensiveBench.

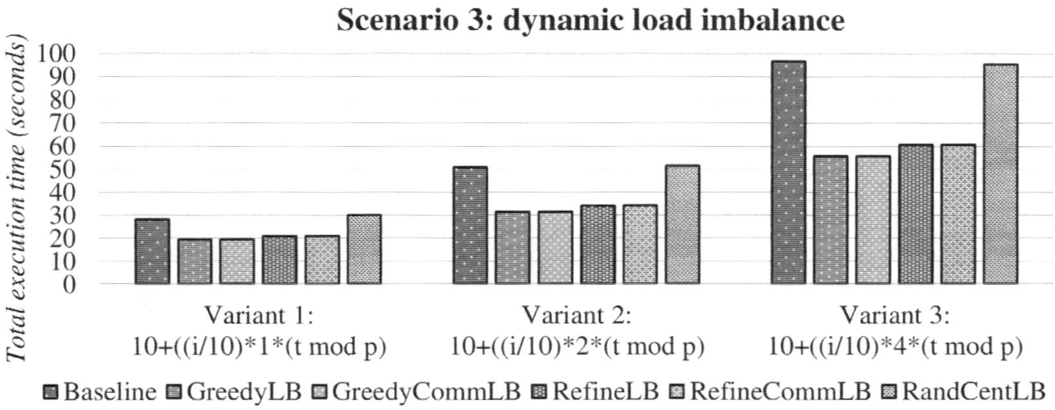

Figure 3. Load balancers' performance with dynamic load imbalance.

Three variants with increasing dynamic load differences were simulated. The total execution time achieved by the different algorithms is shown in Figure 3. In this situation, *GreedyLB* and *GreedyCommLB* improved performance the most. They have shown speedups of 1.7 over the baseline and 1.09 over *RefineLB* and *RefineCommLB* for Variant 3. As these greedy algorithms do no take the current mapping into account, they happen to redistribute tasks before their dynamic behavior takes place. This results in a less extreme load imbalance after the first ten iterations, and a smaller total execution time in the end. Meanwhile, the refinement-based algorithms move almost no tasks during the initial iterations and are faced with more load imbalance due to load dynamicity.

4.4.4. Influence of Communication The fourth and last scenario simulates communication-bound balanced applications. As tasks communicate in a ring and a compact task distribution was applied, locality is maximized with the initial mapping. In this situation, task migrations are unlikely to improve iteration time.

The total execution time achieved by the different scheduling algorithms in this scenario is presented in Figure 4. As the number of messages exchanged between tasks at each iteration increases, so does the total execution time of ComprehensiveBench. No load balancing algorithms was able to improve performance, just as happened in the first tested scenario. However, the increase in execution time in this scenario is a result of slower communication, not migration overhead. Finally, it can be noticed that the communication-aware scheduling algorithms, *GreedyCommLB* and *RefineCommLB*, show no improvement over their communication-oblivious counterparts. These algorithms underestimate the communication costs in this platform because they have no knowledge of the machine topology and memory hierarchy, which results in migrations that affect performance.

4.5. Summary

One of the most important results of all scenarios simulated with ComprehensiveBench is that no single scheduling algorithm was the best in all cases. For instance, *RefineLB* achieved the

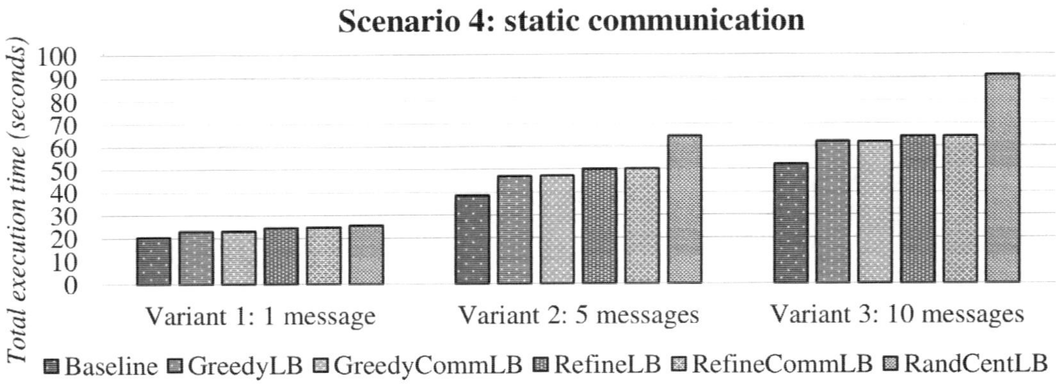

Figure 4. Load balancers' performance with varying numbers of messages.

best results on balanced applications whose tasks have nontrivial memory foot prints (Scenario 1) whereas *GreedyLB* achieved the best results on applications influenced by dynamic load imbalance (Scenario 3). Moreover, both scheduling algorithms presented similar results for imbalanced applications whose tasks have static loads (Scenario 2). Theses results emphasize the most indicated uses for some scheduling algorithms and the scenarios where they should be avoided. They also highlight the importance of a comprehensive benchmark to evaluate the pros and cons of different global scheduling algorithms.

5. Conclusion and Future Work

The difficulty of evaluating and comparing global scheduling algorithms led us to the development of a flexible and expressive benchmark. ComprehensiveBench considers the main characteristics of parallel applications that affect the performance of these algorithms, and enables the simulation of dynamic behavior through parameters set at compile time. ComprehensiveBench helps in the comparison of different global scheduling algorithms and the simulation of parallel applications.

We implemented ComprehensiveBench using the Charm++ parallel programming environment and used it to evaluate load balancing algorithms with four different scenarios. The first scenario highlighted the migration overhead of tasks with large memory footprints by algorithms that do not take the current task mapping into account for their decisions. The second scenario showed how greedy and refinement-based algorithms are able to achieve almost optimal work distributions when handling static load imbalance, while the third scenario illustrated how the greedy algorithms were better suited for dynamic load imbalance. Finally, the fourth scenario highlighted the weakness of *GreedyCommLB* and *RefineCommLB* in estimating the communication costs on NUMA platforms. In all tested cases, ComprehensiveBench allowed us to distinguish between the global scheduling algorithms and showed that no algorithm was the best for all scenarios.

As future work, we plan to employ ComprehensiveBench to evaluate different algorithms on a wide range of parallel platforms. We also plan to develop a mechanism for choosing global scheduling algorithms automatically given the characteristics of an application and a platform. ComprehensiveBench will help by providing varied test cases and enabling the use of machine learning. Finally, we project the use of ComprehensiveBench to emulate other parallel applications and evaluate their migration to environments that support task migration, such as Charm++.

References

[1] Casavant T L and Kuhl J G 1988 *IEEE Trans. Softw. Eng.* **14** 141–154 URL http://dx.doi.org/10.1109/32.4634

[2] Leung J Y T 2004 *Handbook of scheduling: algorithms, models, and performance analysis* Chapman & Hall/CRC computer and information science series (Chapman & Hall/CRC)

[3] Catalyurek U V, Boman E G, Devine K D, Bozdag D, Heaphy R and Riesen L A 2007 *Parallel and Distributed Processing Symposium, 2007. IPDPS 2007. IEEE International* pp 1–11 URL http://dx.doi.org/10.1109/IPDPS.2007.370258

[4] Bhatele A, Kale L V and Kumar S 2009 *Proceedings of the 23rd international Conference on Supercomputing (ICS 2009)* (New York, NY, USA: ACM) pp 110–116 URL http://doi.acm.org/10.1145/1542275.1542295

[5] Rodrigues E R, Navaux P O A, Panetta J, Fazenda A, Mendes C L and Kale L V 2010 *Computer Architecture and High Performance Computing, Symposium on* **0** 71–78

[6] Zheng G, Bhatele A, Meneses E and Kale L V 2011 *International Journal of High Performance Computing Applications (IJHPCA)*

[7] Frasca M, Madduri K and Raghavan P 2012 *Proceedings of the International Conference on High Performance Computing, Networking, Storage and Analysis* SC '12 (Los Alamitos, CA, USA: IEEE Computer Society Press) URL http://dl.acm.org/citation.cfm?id=2388996.2389125

[8] Pilla L L, Ribeiro C P, Cordeiro D, Mei C, Bhatele A, Navaux, Broquedis F, Mehaut J and Kale L V 2012 *Parallel Processing (ICPP), 2012 41st International Conference on* pp 118–127 URL http://dx.doi.org/10.1109/ICPP.2012.9

[9] Menon H and Kalé L 2013 *Proceedings of SC13: International Conference for High Performance Computing, Networking, Storage and Analysis* SC '13 (New York, NY, USA: ACM) URL http://doi.acm.org/10.1145/2503210.2503284

[10] Pilla L L, Ribeiro C P, Coucheney P, Broquedis F, Gaujal B, Navaux P O A and Méhaut J F 2014 *Future Generation Computer Systems* **30** 191–201 ISSN 0167-739X URL http://dx.doi.org/10.1016/j.future.2013.06.023

[11] Tessier F, Mercier G and Jeannot E 2014 *IEEE Transactions on Parallel and Distributed Systems* **25** 993–1002 URL http://dx.doi.org/http://doi.ieeecomputersociety.org/10.1109/TPDS.2013.104

[12] Cruz E H M, Diener M, Pilla L L and Navaux P O A 2015 *23rd Euromicro International Conference on Parallel, Distributed, and Network-Based Processing* pp 207–214

[13] Bailey D, Harris T, Saphir W, Wijngaart R V, Woo A and Yarrow M 1995 The NAS Parallel Benchmarks 2.0

[14] Van der Wijngaart R F and Jin H 2003 NAS Parallel Benchmarks, Multi-Zone Versions

[15] Bienia C, Kumar S, Singh J P and Li K 2008 *Parallel Architectures and Compilation Techniques (PACT)*

[16] Luszczek P R, Bailey D H, Dongarra J J, Kepner J, Lucas R F, Rabenseifner R and Takahashi D 2006 *Proceedings of the 2006 ACM/IEEE Conference on Supercomputing* SC '06 (New York, NY, USA: ACM) URL http://doi.acm.org/10.1145/1188455.1188677

[17] Che S, Sheaffer J W, Boyer M, Szafaryn L G, Wang L and Skadron K 2010 *Workload Characterization (IISWC), 2010 IEEE International Symposium on* pp 1–11

[18] Heroux M A, Doerfler D W, Crozier P S, Willenbring J M, Edwards H C, Williams A, Rajan M, Keiter E R, Thornquist H K and Numrich R W 2009 *Sandia National Laboratories, Tech. Rep. SAND2009-5574*

[19] Tesser R, Pilla L L, Navaux P O A, Dupros F, Mehaut J F and Mendes C 2014 *To be published on Parallel, Distributed and Network-Based Processing (PDP), 2014 22st Euromicro International Conference on* pp 1–8

[20] Lifflander J, Krishnamoorthy S and Kale L V 2012 *Proceedings of the 21st international symposium on High-Performance Parallel and Distributed Computing* HPDC '12 (New York, NY, USA: ACM) URL http://doi.acm.org/10.1145/2287076.2287103

[21] Touati S A A, Worms J and Briais S 2013 *Concurrency and Computation: Practice and Experience* **25** 1410–1426 URL http://dx.doi.org/10.1002/cpe.2939

[22] Kale L V and Krishnan S 1993 *Proceedings of the Eighth Annual Conference on Object-Oriented Programming Systems, Languages, and Applications (OOPSLA 1993)* (ACM) pp 91–108

[23] Acun B, Gupta A, Jain N, Langer A, Menon H, Mikida E, Ni X, Robson M, Sun Y, Totoni E, Wesolowski L and Kale L 2014 *High Performance Computing, Networking, Storage and Analysis, SC14: International Conference for* pp 647–658 URL http://dx.doi.org/10.1109/SC.2014.58